Bitcoin and Blockchain

Internet of Everything (IoE)

Series Editor:
Mangey Ram
Professor, Graphic Era University, Uttarakhand, India

IoT
Security and Privacy Paradigm
Edited by
Souvik Pal, Vicente Garcia Diaz, and Dac-Nhuong Le

Smart Innovation of Web of Things
Edited by
Vijender Kumar Solanki, Raghvendra Kumar, and Le Hoang Son

Big Data, IoT, and Machine Learning
Tools and Applications
Rashmi Agrawal, Marcin Paprzycki, and Neha Gupta

Internet of Everything and Big Data
Major Challenges in Smart Cities
Edited by
Salah-ddine Krit, Mohamed Elhoseny, Valentina Emilia Balas, Rachid Benlamri, and Marius M. Balas

Bitcoin and Blockchain
History and Current Applications
Edited by
Sandeep Kumar Panda, Ahmed A. Elngar, Valentina Emilia Balas, and Mohammed Kayed

For more information about this series, please visit: https://www.crcpress.com/Internet-of-Everything-IoE-Security-and-Privacy-Paradigm/book-series/CRCIOESPP

Bitcoin and Blockchain
History and Current Applications

Edited by
Sandeep Kumar Panda, Ahmed A. Elngar,
Valentina Emilia Balas, and Mohammed Kayed

CRC Press
Taylor & Francis Group
Boca Raton London New York

CRC Press is an imprint of the
Taylor & Francis Group, an **informa** business

MATLAB® is a trademark of The MathWorks, Inc. and is used with permission. The MathWorks does not warrant the accuracy of the text or exercises in this book. This book's use or discussion of MATLAB® software or related products does not constitute endorsement or sponsorship by The MathWorks of a particular pedagogical approach or particular use of the MATLAB® software.

First edition published 2020
by CRC Press
6000 Broken Sound Parkway NW, Suite 300, Boca Raton, FL 33487-2742

and by CRC Press
2 Park Square, Milton Park, Abingdon, Oxon, OX14 4RN

ISBN: 978-0-367-90100-4 (hbk)
ISBN: 978-1-003-03258-8 (ebk)

Typeset in Times
by codeMantra

Dedicated to my sisters Sujata and Bhaina Sukanta, nephew Surya Datta, wife Itishree (Leena), my son Jay Jagdish, and my late father Jaya Gopal Panda and late mother Pranati Panda.

Sandeep Kumar Panda

Dedicated to my parents, my brother, my sisters, and also to my kids Farida and Seif, your smile brings happiness in my life.

Ahmed A. Elngar

Contents

Preface

Globally, the industries provide employment to about 500 million people from the main business sectors which include service, retail, manufacturing, business, health care, local and central government, finance sector, etc. All of these sectors are made either automatic or semiautomatic by sophisticated business processes forming an integral part of the digital economy. In this revolution, the Internet plays a vital role in core business, and financial aspects of the digital economy are still centralized, with the help of centralized agencies such as banks and tax agencies, to authenticate and settle payments and transactions. These centralized services often are manual, difficult to automate, and represent a bottleneck to facilitating a frictionless digital economy. The blockchain technology, a distributed, decentralized, and public ledger, addresses these issues by maintaining records of all transactions on a blockchain network that promises a smart world of automation of complex services and manufacturing processes.

A blockchain network is a peer-to-peer network and does not require a central authority or trusted intermediaries to authenticate or settle the transactions. The first generation of blockchain network was coined by Satoshi Nakamoto in his 2008 white paper where the primary application of the blockchain network was the use of electronic cash or cryptocurrency called Bitcoin. The second-generation blockchain network called Ethereum was introduced in 2013. Ethereum allows a single programmable blockchain network to be used for developing different types of applications where each application takes the form of a smart contract which is implemented in a high-level language and deployed on the blockchain network.

There are very few books that can serve as a foundational text book for colleges and universities looking to create new educational programs in the areas of blockchain, smart contracts, and decentralized applications. The existing books are focused on the business side of blockchain and case study–based evaluation of applications.

We have edited this book to meet the need at colleges and universities. This book can serve as a textbook for senior- and graduate-level courses on the domains such as business analytics, finance, Internet of Things, computer science, mathematics, and business schools. This book is also dedicated to novice programmers and solution architects who want to build powerful, robust, optimized smart contracts using solidity, and hyperledger fabrics from scratch.

This book is organized into 13 chapters.

Chapter 1, *Bitcoin: A P2P Digital Currency*, presents an overview of digital currencies, discusses the Bitcoin transactions and proof of work, reviews the Bitcoin security attacks, and illustrates the Bitcoin development environment.

Chapter 2, *Exploring the Bitcoin Network*, describes the overall process of Bitcoin networks like transactions, digital signatures, relay networks, and Bitcoin script.

Chapter 3, *Blockchain Technology: The Trust-Free Systems*, discusses blockchain technology in detail and highlights some of the applications in which it can be used. It also specifies some challenges and benefits of this technology that is all set to transform the digital world.

Chapter 4, *Consensus and Mining in a Nutshell*, provides an overview of the consensus models, narrates the transaction process using consensus, and examines the different consensus attacks.

Chapter 5, *Blockchain: Introduction to the Technology behind Shared Information*, discusses the incentive to the miner, attacks, types of blockchain, and blockchain impacts in finance and industry sectors.

Chapter 6, *Growth of Financial Transaction toward Bitcoin and Blockchain Technology*, deliberates the introduction to cryptocurrency, and its history and definitions. Nevertheless, this chapter also discusses some cybersecurity aspects of blockchain.

Chapter 7, *A Brief Overview of Blockchain Algorithm and Its Impact upon Cloud-Connected Environment*, focuses on blockchain algorithm and its impact upon cloud-connected ecosystem.

The focus of Chapter 8, *Solidity Essentials*, is on briefly understood Solidity Language. The need for Solidity is discussed. Its use case and implementation are addressed. Details of its environment setup and compilation are also provided. Its important components are explained along with examples for a better understanding of syntax. By the end of this chapter, the reader will be familiar with Solidity and will be able to write smart contracts on it.

Chapter 9, *Installing Frameworks, Deploying, and Testing Smart Contracts in Ethereum Platform*, covers the smart contract programming with a special emphasis on Solidity, properties associated with a smart contract account, fetching accounts from Ganache module, deployment of smart contracts with Web3 and Infura, testing smart contracts with open-source tools such as Remix and Mocha framework for asynchronous testing, test coverage reports, and use of any assertion library. Hence, this chapter highlights the methods for writing smart contracts using various authoring tools such as Visual Studio. However, the easiest and fastest way for developing and testing the smart contracts is to use a browser-based tool known as Remix. Next, we write few smart contract applications with Solidity language and explain all the common function types used in it. The concepts of gas and transactions are thoroughly discussed. Lastly, this chapter is laid out in a manner that it helps the reader to get the comprehensive idea of writing smart contracts.

Chapter 10, *Blockchain in Healthcare Sector*, discusses the existing and the latest new developments in implementation, challenges, applications, and future direction of the blockchain-based systems in healthcare sectors.

Chapter 11, *Blockchain Theories and Its Applications*, describes the financial applications where blockchain provides traceability and transparency to the financial transactions, which makes it a unique technology to take care of financial applications. Financial applications in the area of banking services, insurance sector – health insurance, economic business applications, financial auditing, and cryptocurrency payment and exchange are part of discussion. Non-financial applications: health care – in healthcare sector, patients' records should be shared with the healthcare stakeholders with confidentiality; blockchain is an effective technology to take care of this process. Governance – blockchain technology can serve as a path changer for the local and central governments to take care of governance.

Chapter 12, *Building Permissioned Blockchain Networks Using Hyperledger Fabric*, discusses in detail about Hyperledger Fabric, its architecture, and the procedure to build the Hyperledger Fabric network using docker container technology for an industry use case, which would involve multiple actors/organizations who want to form a permissioned network. We will also discuss about developing the chaincode (smart contract) for Hyperledger Fabric using node js, and building a Representational State Transfer Application Program Interface (REST API) to interact with the Fabric network using Fabric Node Software Development Kit (SDK).

Chapter 13, *Fraud-Resistant Crowdfunding System Using Ethereum Blockchain*, presents all the details and technicalities in implementation of a crowdfunding platform through Ethereum blockchain network in a scholarly manner.

Editors

Dr. Sandeep Kumar Panda is currently working as an associate professor in the Department of Computer Science and Engineering, Faculty of Science and Technology at ICFAI Foundation for Higher Education (deemed to be University), Hyderabad, Telangana, India. His research interests include Software Engineering, Web Engineering, Cryptography and Security, Blockchain Technology, Internet of Things, and Cloud Computing. He has published many papers in international journals and international conferences in repute. He received the "Research and Innovation of the Year Award" hosted by WIEF and EduSkills under the Banner of MSME, Government of India and DST, Government of India at New Delhi in January 2020. He has eight Indian Patents in his credit. His professional affiliations are MIEEE, MACM, and LMIAENG.

Dr. Ahmed A. Elngar is an assistant professor of Computer Science, the founder and chair of Scientific Innovation Research Group (SIRG), and the Director of Technological and Informatics Studies Center, Faculty of Computers & Artificial Intelligence at Beni-Suef University, Egypt. He is managing editor of *Journal of Cybersecurity and Information Management* (JCIM). He has published more than 25 scientific research papers in prestigious international journals and over five books covering such diverse topics as data mining, intelligent systems, social networks, and smart environment. Research works and publications. He is a collaborative researcher. He is a member of the Egyptian Mathematical Society (EMS) and International Rough Set Society (IRSS). His other research areas include Internet of Things (IoT), Network Security, Intrusion Detection, Machine Learning, Data Mining, Artificial Intelligence, Big Data, Authentication, Cryptology, Healthcare Systems, and Automation Systems. He is an Editor and Reviewer of many international journals. He won several awards including the "Young Researcher in Computer Science Engineering" from Global Outreach Education Summit and Awards 2019 on 31 January 2019 (Thursday) at Delhi, India and the "Best Young Researcher Award (Male) (Below 40 Years)" from Global Education and Corporate Leadership Awards (GECL-2018) at

Rajasthan, India. Also, he has an Intellectual Property Rights called "ElDahshan Authentication Protocol," Information Technology Industry Development Agency (ITIDA), Technical Report, 2016. His activities in community and the environment service include organizing 12 workshops hosted by a large number of universities in almost all governorates of Egypt. He has participated in a workshop on Smartphone's techniques and their role in the development of visually impaired skills in various walks of life.

Prof. Valentina Emilia Balas is currently a full professor in the Department of Automatics and Applied Software at the Faculty of Engineering, "Aurel Vlaicu" University of Arad, Romania. She holds a Ph.D. in Applied Electronics and Telecommunications from Polytechnic University of Timisoara. She is the author of more than 300 research papers in refereed journals and international conferences. Her research interests are in Intelligent Systems, Fuzzy Control, Soft Computing, Smart Sensors, Information Fusion, Modeling, and Simulation. She is the editor-in-chief of *International Journal of Advanced Intelligence Paradigms* (IJAIP) and *International Journal of Computational Systems Engineering* (IJCSysE), is a member of Editorial Board of several national and international journals, and is expert evaluator for national and international projects and Ph.D. theses. She is the director of Intelligent Systems Research Centre and director of the Department of International Relations, Programs and Projects at "Aurel Vlaicu" University of Arad. She served as the general chair of the International Workshop Soft Computing and Applications (SOFA) in eight editions held from 2005 to 2018 in Romania and Hungary. She participated in many international conferences as organizer, honorary chair, session chair, and member in Steering, Advisory, or International Program Committees. She is a member of EUSFLAT, SIAM, TC – Fuzzy Systems (IEEE CIS), TC – Emergent Technologies (IEEE CIS), and TC – Soft Computing (IEEE SMCS), and a senior member of IEEE. She was past vice-president (Awards) of IFSA International Fuzzy Systems Association Council (2013–2015) and is a joint secretary of the Governing Council of Forum for Interdisciplinary Mathematics (FIM), a multidisciplinary academic body, India.

Dr. Mohammed Kayed received an M.Sc. degree in Computer Science from Minia University, Minia, Egypt in 2002 and a Ph.D. degree in Computer Science from Beni-Suef University, Beni-Suef, Egypt in 2007. From 2005 to 2006, he was a research and teaching assistant in the Department of Computer Science and Information Engineering at the National Central University, Taiwan. Since 2007, he has been an assistant professor with Department of Mathematics and Computer Science, Faculty of Science, Beni-Suef University, Beni-Suef, Egypt. He is currently an associate professor and head of Computer Science Department, Faculty of Computer and Artificial Intelligence, Beni-Suef University, Egypt. He is the author of more than 18 articles. His research interests include Web Mining, Opinion Mining, Information Extraction, and Information Retrieval.

Contributors

Naseem Ahamed
Finance and Accounting Deapartment
ICFAI Business School
Hyderabad, Telangana, India

Chiranji Lal Chowdhary
School of Information Technology and
   Engineering
VIT University
Vellore, Tamil Nadu, India

Jaipal Dhobale
Operation and IT Deapartment
ICFAI Business School
Hyderabad, Telangana, India

Ahmed A. Elngar
Computer Science
Beni-Suef University
Beni-Suef, Egypt

Parv Garg
Computer Science and Engineering
ICFAI Foundation for Higher
   Education (Deemed to be
   University)
Hyderabad, Telangana, India

Ajay Kumar Jena
School of Computer Engineering
KIIT (Deemed to be University)
Bhubaneswar, Odisha, India

D. Kesavaraja
Department of Computer Science and
   Engineering
Dr. SivanthiAditanar College of
   Engineering
Tiruchendur, Tamil Nadu, India

Neeraj Khadse
Computer Science and Engineering
ICFAI Foundation for Higher Education
   (Deemed to be University)
Hyderabad, Telangana, India

Vaibhav Mishra
Operation and IT Deapartment
ICFAI Business School
Hyderabad, Telangana, India

Subhasish Mohapatra
School of Computer Engineering
ADAMAS University
Kolkata, West Bengal, India

Sandeep Kumar Panda
Computer Science and Engineering
ICFAI Foundation for Higher Education
   (Deemed to be University)
Hyderabad, Telangana, India

Smita Parija
Electrical Engineering Department
National Institute of Technology
Rourkela, Odisha, India

S. Porkodi
Department of Computer Science and
   Engineering
Dr. SivanthiAditanar College of
   Engineering
Tiruchendur, Tamil Nadu, India

K. Varaprasada Rao
Computer Science and Engineering
ICFAI Foundation for Higher Education
   (Deemed to be University)
Hyderabad, Telangana, India

P. Praneeth Reddy
Computer Science and Engineering
ICFAI Foundation for Higher Education
    (Deemed to be University)
Hyderabad, Telangana, India

S. Saikrishna
Computer Science and Engineering
ICFAI Foundation for Higher Education
    (Deemed to be University)
Hyderabad, Telangana, India

Sathya A.R.
Computer Science and Engineering
ICFAI Foundation for Higher Education
    (Deemed to be University)
Hyderabad, Telangana, India

Tushar Sharma
Computer Science and Engineering
ICFAI Foundation for Higher Education
    (Deemed to be University)
Hyderabad, Telangana, India

Santosh Kumar Swain
School of Computer Engineering
KIIT (Deemed to be University)
Bhubaneswar, Odisha, India

Mutyala Sree Teja
Computer Science and Engineering
ICFAI Foundation for Higher Education
    (Deemed to be University)
Hyderabad, Telangana, India

1 Bitcoin
A P2P Digital Currency

Sathya A.R.
ICFAI Foundation for Higher Education
(Deemed to be University)

Ahmed A. Elngar
Beni-Suef University

CONTENTS

1.1   INTRODUCTION

Over the past few decades, there have been many new applications on the Internet, solving problems in a supportive and distributed way. There are many well-known applications that are non-commercial and collaborative; for example, Hashcash, Anonymous Communication, and BitTorrent. Generally, few applications materialize soon after the software concept is perceived. But, there are few exceptions; one such idea is digital money. The concept of digital money is around since the 1980s, but it took few decades to develop as a fully distributed solution. As per Ref. [1], several attempts had been made to build digital currencies, but they required a central authority like bank to handle transactions. Later approaches such as Bitgold, B-money, Reusable Proof of Work (RPoW), and Karma suggested to have a cryptographic puzzle named proof of work (PoW). Based on this model, every user can mine their money but using a central bank to maintain the transactional details.

To eliminate the central authority, the register which records all transaction details should also be distributive in nature. However, a serious risk involved in digital currencies or any distributed currency is double spending. As digital copies are smaller, it is possible for any malicious user to make two parallel transactions using the same coins to two different users. In traditional centralized models, the central authority like banks can identify and prevent such malicious acts, whereas it is not easy in a distributed model. The reason is that keeping the distributed information and mutual acceptance in a consistent state is challenging, especially in a malicious users environment. This goes down similar to the Byzantine Generals Problem [2]. This vision leads to employing quorum systems. Quorum systems [3] understand and accept that in a distributed environment, it is possible to have faulty information and mischievous users. Hence, to ensure a correct ledger, the voting concept was introduced assuming majority of peers are honest. However, this election approach leads to Sybil attacks [4], in which malicious users can set up several peers to challenge the electoral process and implant wrong information. Propagation delays are ignored and lead to brief inconsistencies. Bitcoin design overcomes all these difficulties.

1.2   DIGITAL CURRENCIES BEFORE BITCOIN

One of the earliest attempts to create cryptocurrency started few decades back in the Netherlands. When a petrol station in the Netherlands suffered nighttime thefts, few developers tried to link money with newly designed smart cards. A user who needs to access the petrol station can use these smart cards instead of cash. This can be one of the earliest examples of electronic cash which might have led to the digital currency as we know them today. Some digital cash concepts before Bitcoin are explained below.

1.2.1 BLINDED CASH

In 1983, David Chaum was the first to describe digital money. He states that the key difference between credit-card payment and digital cash is anonymity. According to Chaum, users can anonymously receive digital money from banks. Banks can view who exchanged and how much money got exchanged, but cannot know what it is used for. Chaum uses cryptography to create a blind, digital signature termed "blinded cash" to make the cash anonymous. Therefore, "blinded cash" could be exchanged securely between individuals, by carrying identity signatures and the ability to change without traceability. But in 1988, Chaum went bankrupt. However, his concepts, formula, and encryption tools played a key role in developing digital currencies later.

1.2.2 WEB-BASED MONEY

In the 1990s, many companies tried to enlarge Chaum's ideas. One such company is PayPal. It allowed the users to send money quickly and safely via web browser. Combined with eBay, it secured a dedicated user base and still remains a major payment service. Some companies even tried to trade gold via web browser. One successful operation was e-gold, which offered credit in exchange of gold or other precious metals. But in 2005, the federal government has to shut this down due to scams.

1.2.3 B-MONEY

Wei Dai, a developer, introduced a distributed, anonymous electronic cash system called B-money in 1998. B-money used digital pseudonyms in a distributed environment to transfer digital currency. It even enforces contracts in the network without using a third party. But B-money was not successful and failed to get any attention. However, Satoshi used some elements of B-money in his Bitcoin whitepaper.

1.2.4 BIT GOLD

Later, Bitgold, which is another digital currency, was proposed by Nick Szabo. Similar to today's Bitcoin mining process, Bitgold had its own PoW mechanism through which the solutions were cryptographically signed and broadcasted to the public. Bitgold to an extent tried to avoid dependency on central authorities. However, it was not successful, but added bits to the evolution of digital currencies.

1.2.5 HASHCASH

According to the Merkle, Hashcash is one of the most successful pre-Bitcoin digital currencies developed in the mid-1990s. Hashcash was developed to serve many purposes like reduce email spams, prevent Distributed Denial-of-Services (DDoS) attacks, etc. Like many other modern digital currencies, Hashcash also uses PoW algorithm to help in generating and distributing new coins. In fact, in 1997, Hashcash also ran into the same issues faced by the other currencies. Hashcash eventually became less effective because of its increased processing power requirement. In 2009, when Bitcoin was proposed, it started a new generation of digital currencies. Bitcoin's association with blockchain technology and decentralization status make it different from many of the cryptocurrencies developed before. At the same time, it is not possible to visualize Bitcoin, leaving the earlier attempts at cryptocurrencies.

1.3 BITCOIN IN A NUTSHELL

Bitcoin and many other cryptocurrencies use open-source software to solve problems related to peer-to-peer (P2P) network. A P2P system functions in a way different from a government-issued fiat currency. Creating, certifying, and issuing fiat money is performed by one person and is used by many. It is more like a client server model in networking where a server receives and responds to requests from many clients. A server is responsible for ensuring the correctness of data or information provided to it. Making fake fiat currency is difficult and is an offense.

A group of nodes that are linked to a network is called a P2P network. The nodes act both as client and as server. P2P network is faster, and its cost of maintenance is lesser than the client server model. It is more flexible to attacks or issues at one location. In addition to relying on a P2P network, Bitcoin also depends on the open-source software. In open-source software, the source code is distributed with no or less limitation on copyright in using the program.

Bitcoin design overcomes the difficulties faced by the digital currencies. It was announced and deployed in 2008 and 2009, respectively, by Satoshi Nakamoto. Bitcoin gained popularity immediately after that. The identity of Nakamoto remains uncertain and leads to many speculations. Bitcoin brilliantly combines many existing technologies [5,6] and makes decentralization practical by limiting the number of votes per entity in PoW scheme.

All transactions will be collected in a block by Bitcoin miners and the miner will attempt to solve the given cryptographic puzzle by changing its nonce value. Once the solution is obtained, it can be broadcasted along with the transactions to other nodes and can update the distributed ledger. Coin ownership can be found by traversing the blockchain until the coin's most recent transaction is found. Due to malicious operations and transmission delays, it is possible to have forks in the chain. Consensus on such cases can be reached by taking the longest (block) chain for consideration. In this way, Sybil and double spending attacks can be mitigated to an extent by adding blockchain to PoW contributions. PoW generates the supply of Bitcoin continuously for miners as incentives. Bitcoin does not require a central authority to do these activities, thus making the distributed, digital currency practical. Table 1.1 differentiates the various features of centralized trust-based models and decentralized model.

TABLE 1.1

Comparison of Centralized and Decentralized Models

Features	Centralized Model	Decentralized Model (Bitcoin)
Transaction guidelines	Central authority	Consensus
Transaction verification	Central authority	Consensus
Money generation	Through loan	Through mining
Money supply	Unlimited	Limited
Money value	Based on exchange rate	Based on PoW, supply, and demand
Money transfer	Reversible with central authority	No central authority. Direct and non-reversible
Privacy	Partially anonymous, but known to central authority	Partially anonymous
Fee	Transaction fee	Considerably low transaction fee
Transaction delay	Delay in days	Delay in minutes

1.4 TRANSACTION

A publicly available database records every cryptocurrency exchange. Every Bitcoin has an address called Bitcoin address. Transfer of Bitcoins from one Bitcoin address to another is a transaction. Transactions that are yet to be recorded and most recent transactions are stored in blocks. Every time a block is completed, it'll bring a new block into picture. Block generation produces rewards to the users. The reward system implements a new, unique transaction in the block as the first transaction. This first transaction is labeled as Coinbase transaction. The miner is always the creator of that block. A transaction is never completed until it gets added in a block. The miners compete to add the block to the existing blocks. The process of adding a new block is called mining. Every block comes with an incentive. New Bitcoins are generated and issued to the creator of the block at each block generation. Transactions are hashed, combined, and hashed once again at each block until a single hash value is obtained. This is called Merkle root [7]. Merkle root and the hash value of the previous block are stored in the block header. A typical Bitcoin transaction structure is given in Figure 1.1.

1.4.1 CONSTRUCT A TRANSACTION

User's wallet software will have all logic to choose a required input and output for a transaction. Once the user selects the Bitcoins to be transferred and the destination address, rest of the process will be taken care by the wallet itself. It is not necessary for a user to be online for the wallet to construct a transaction. It can construct even if the user is offline; the user only needs to be connected to the network to execute a transaction.

1.4.2 GETTING THE RIGHT INPUT

To begin with, the balance of the user has to be verified by the wallet. This ensures that the user has sufficient balance to make a transaction. For this reason, every

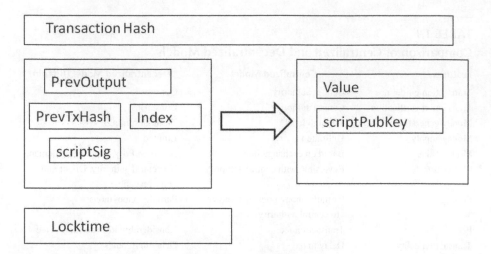

FIGURE 1.1   Bitcoin transaction.

wallet maintains the balance of unspent transaction output. If no such copies are maintained in the wallet, it can always request the network to retrieve the information. Various application programming interfaces (APIs) are available to do this task. Once the required amount of Bitcoin is available in the wallet, it can create the output of the transaction.

1.4.3   CREATING THE OUTPUT

A script is created for the transaction output by which the transaction values can be redeemed by providing a solution to the script; that is, whoever provides the signature from the key confirming to the recipient's public address. As the destination wallet will have the key with respect to that address only, the intended recipient can provide the signature required and can redeem the coins. At last, for the transaction to happen in a proper way, wallet application of the sender will charge a small fee. This charging will not be explicit; it is the difference between the input and the output. This transaction fee will later be collected by the miner to validate and add a block in the network.

To manage the input and output, Bitcoin uses a scripting language which imposes the conditions to be met to redeem the Bitcoins. The most popular script is "Pay-to-PubkeyHash" (P2PKH). It needs only one signature of the owner to approve the payment. In contrast, another script "Pay-to-ScriptHash" uses a multisignature addresses for various transactions.

The schematic representation of a Bitcoin transaction is shown in Figure 1.2.

1.5   TIMESTAMP SERVER

The timestamp server functions by taking hash of a block comprising transactions and distributing them to the public like newspapers or Usenet. The purpose of

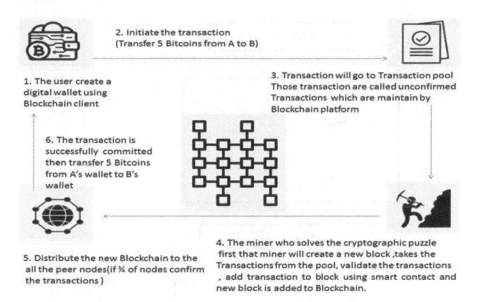

2. Initiate the transaction
(Transfer 5 Bitcoins from A to B)

1. The user create a
digital wallet using
Blockchain client

3. Transaction will go to Transaction pool
Those transaction are called unconfirmed
Transactions which are maintain by
Blockchain platform

6. The transaction is
successfully committed
then transfer 5 Bitcoins
from A's wallet to B's
wallet

5. Distribute the new Blockchain to the
all the peer nodes(if ¾ of nodes confirm
the transactions)

4. The miner who solves the cryptographic puzzle
first that miner will create a new block ,takes the
Transactions from the pool, validate the transactions
, add transaction to block using smart contact and
new block is added to Blockchain.

FIGURE 1.2 Bitcoin transaction summary.

timestamp server is to guarantee that a block of transactions actually exist at the time of timestamp. This allows the nodes to verify the order in which the transactions are distributed. Hence, it is possible for the users to have a history of all transactions over the network. Figure 1.3 shows the process of time stamping.

1.6 PROOF OF WORK

In Bitcoin, multiple copies of blockchain exist in the network, and it is necessary to keep the global view of the blockchain consistent. For example, two different transactions can be created using the same coins to different receivers, which is called "double spending". And if the two receivers process and get their transactions verified individually based on their local views of blockchain, the blockchain becomes inconsistent. Such problems can be resolved by (i) sharing verification process of the transaction to guarantee rightness of the transaction and (ii) letting everyone know the successfully processed transaction to assure the consistency of blockchain. Bitcoin uses PoW and a distributed consensus protocol to satisfy the above specifications.

As mentioned, the distribution of verification process is to ensure the correctness of data by majority of valid users. Therefore, if the blockchain goes into an unstable state at any time, all users can update their local blockchain copy to the one accepted by most miners. But this election process is subject to Sybil attacks [4]. In Sybil attacks, a malicious user creates many virtual nodes and makes them vote for a wrong transaction, thereby disturbing the election process.

PoW model helps Bitcoin to overcome the Sybil attacks. In this model, a miner has to solve a cryptographic mathematical puzzle to prove their genuineness similar

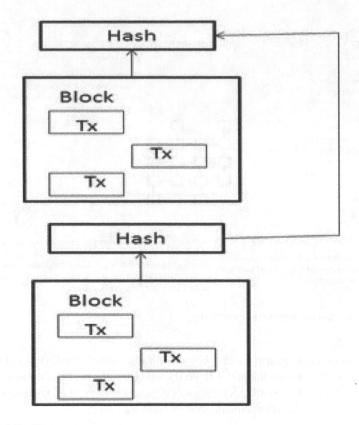

FIGURE 1.3 Timestamp server.

to Hashcash. PoW includes a value named "nonce" which generates a hash value when hashed with SHA-256 which starts with the required number of zero bits. The target sets the needed number of zero bits. The hash value needed must be below the current target value and should have a certain number of leading zero bits to be less than the target. To do this, a high level of computational cost is applied to verification by PoW, and to do PoW, the miner needs a high level of computing resources. It is, therefore, very difficult to fake the computing resources. Thus, the problem of Sybil attacks is resolved.

In general, the transactions are not mined individually. The miners collect the pending transactions to make a block and then mine the block by measuring the block's hash value as well as varying nonce. The nonce must differ until the final value is less than or equal to the given target value. The miners share the 256-bit target value. It is not simple to calculate the hash value. SHA-256 is the hash function used in Bitcoin [8]. If the cryptographic hash function does not find the necessary hash value, the only alternative is to try different nonce until a solution is found (a hash value below the target). Consequently, the complexity of the puzzle depends on the target value, i.e., lower the target value, lesser the number of solutions, thereby making it more difficult to calculate the hash. Once the correct hash value is determined for a block, the miner immediately broadcasts the block to the network along with the measured hash value and nonce, and then the block is appended to his private blockchain. On receiving a mined block, the other miners compare the hash value given to the target value in the received block and verify its correctness. They will also upgrade their blockchain locally by adding the new block. After adding the block in blockchain successfully, the first who solved the puzzle will get a reward. There is no central authority to provide the rewards. In the block generation system, whenever a miner adds a Coinbase transaction or a reward-generating transaction for his Bitcoin address, the rewards are provided.

Other than these rewards, the miner also receives a transaction fee for every successful addition of transaction in block. The transaction fee usually will be the difference of value of all inputs and values of all outputs in a transaction. Studies show that higher transaction fees make the transactions with lower transaction fees to suffer from starvation problem, i.e., service denied for a longer duration. Bitcoin never necessitates transaction fee. It is the owner of the transaction who sets the fee and is not a constant value. Nonetheless, as users struggle to get transactions publicized on the blockchain, transaction fees increase to rates that discourage the use of Bitcoin, illustrating a major structural issue facing the blockchain.

1.7 BITCOIN DEVELOPMENT ENVIRONMENT

Bitcoin is an open-source mission, and MIT holds the license for its source code. Bitcoin Core can be downloaded for free. A community of volunteers had created the Bitcoin project. Initially, only Satoshi Nakamoto was there in the project. Later by 2016, it had around 400 contributors with a handful of developers and several part-time developers. Bitcoin Core can be regarded as the reference implementation of the Bitcoin system by Bitcoin since it is the authoritative reference on how to implement each part of the technology. It is labeled as the Satoshi client. Bitcoin Core acts as a Bitcoin node, and a collection of Bitcoin nodes form a Bitcoin network. Nakamoto initially released the "Bitcoin" software, but to make it different from the network, he later renamed it to "Bitcoin Core." Bitcoin Core has all features of Bitcoins such as wallet, block validation engine, transaction, and full network node in P2P Bitcoin network.

1.7.1 BITCOIN CORE IMPLEMENTATION

Bitcoin Core can be downloaded from http://Bitcoin.org/en/choose-your-wallet and installed by clicking Bitcoin Core button. Once the installation is complete, a new application named "Bitcoin–QT" will be listed in applications. Bitcoin client can be started by double clicking the icon. As the Bitcoin client starts running, it will download the blockchain, and it will take several days to complete. When "Synchronized" message is displayed, it is understood that the downloading process is successful.

The Bitcoin developers can download the full source code either as ZIP file or it can be cloned from the authoritative database https://github.com/Bitcoin/Bitcoin. On the other hand, in your system, a local copy of the source code can be created using git command line.

The below command in Linux OS is used to clone the source code.

```
$ git clone https://github.com/Bitcoin/Bitcoin.git
```

After execution of the above command, a local copy of the complete database of source code will be available in Bitcoin directory. Documentation will also be offered along with the source code and is available in numerous files. The main documentation will be available in README.md file in Bitcoin directory. Get into the Bitcoin directory by executing the below command.

```
$ cd bitcoin
```

In order to ensure that the system in which the source code is downloaded has all libraries to compile, the following script has to be executed to find the correct setting.

```
$ ./autogen.sh
```

The above autogen.sh script generates a set of automatic configuration scripts. These configuration scripts will examine your system to find the correct settings, and will ensure that the system has all the needed libraries to compile the code.

The most important script is the "Configure" script. It facilitates various options to customize the build process. The command ./configure --help will list the options available to create a customized build for your system.

```
$ ./configure
```

This script will find all the necessary files automatically, and thus a customized build will be created. If everything is well, the configure script will finish its execution without any error and create customized build scripts which will help in compiling Bitcoin. Otherwise, it will terminate with errors. The error possibly could be a missing or incompatible library. In that case, analyze the build document again and be assured that necessary prerequisites are installed, and then rerun the configure code.

The next step is to compile the source code, which will take hours to complete. The compilation process has to be monitored often to check whether any error message is shown.

```
$ make
```

Once the source code is compiled without any errors, installing the Bitcoin executable into the system path is the final step. This can be done by executing the make command.

```
$ sudo make install
```

To confirm the successful installation of Bitcoin, the following commands can be executed. The sudo make install command asks for the path of the two executables.

```
$ which bitcoind
/usr/local/bin/bitcoind

$ which bitcoin-cli
/usr/local/bin/bitcoin-cli
```

Bitcoin's default install is set to be in /usr / local / bin. When we first run the Bitcoin command, we'll be told to create a strong password for the JSON-RPC interface.

```
$ Bitcoind
```

Execution of this command will throw an error asking the user to set rpcpassword in the configuration file. To change the password, the configuration file can be edited and a new password can be set as per the Bitcoin recommendation. At last, run the Bitcoin Core client. While running it for the first time, the Bitcoin blockchain will be rebuilt, and it takes on average two days to complete.

1.7.2   WALLET SETUP AND ENCRYPTION

The next step is to set up the wallet configuration. To start with, we encrypt the wallet with a password.

```
$ Bitcoin-cli encryptwallet newpwd
```

In the above command, the command encryptwallet is used to encrypt the wallet and the newpwd is the password set for the wallet. Similarly, commands like backupwallet, importwallet, and dumpwallet are used to back up, restore, and dump the wallet in text file, respectively.

```
$ Bitcoin-cli backupwallet wallet.backup
$ Bitcoin-cli importwallet wallet.backup
$ Bitcoin-cli dumpwallet wallet.txt
```

After setting up the wallet with a password, it can be configured to receive transactions. A group of Bitcoin addresses are maintained by Bitcoin reference implementation. These addresses are used as receiving address and change address. The addresses are generated automatically. To acquire an address, the command getnewaddress has to be executed.

```
$ Bitcoin-cli getnewaddress
```

Now, a random address would be provided. For instance, the newly generated address can be: 1hvzAofGwT8cjb8JU4nBsCSfEVWX5u9CL. This can be used to receive money from external wallets. To view all the transactions received by the wallet, the following command can be used.

```
$ Bitcoin-cli listtransactions
[
```

```
{
  "account" : "",
  "address" : "1hvzSofGwA8cjb8JU7nBsGSfEVQX5u6CL",
  "category" : "Receive",
  "amount" : 0.06000000,
  "confirmations" : 0,
  "txid" :
"9ca8f969be3ef5ec2a8685660fdbf7g8bd365524c2e1fc66c319
acbae2c14ke3",
  "time" : 1392660908,
  "timereceived" : 1392660908
}
]
```

The "getaddressesbyaccount" command is used to get all the addresses in the wallet.

```
$ Bitcoin-cli getaddressesbyaccount ""
```

Now, information from blocks and transactions can be discussed. For instance, if we know the block in which the transaction is added, we can inquire the block using the "getblock" command.

```
$ Bitcoin-cli getblock
```

The getblock command lists all the transactions that are present in the block. For example, if the command shows an output of say 250 transaction ids, it can be interpreted as 367 transactions are present in the block. listunspent, gettxout, createrawtransaction, and sendrawtransaction are few transaction commands used to see the unspent output, get details of the unspent output, create transaction, and make transaction, respectively.

1.8   KEYS, WALLETS, AND ADDRESSES

1.8.1   DIGITAL SIGNATURE

Every user in the network has a pair of private and public keys. The private key is used to sign the transactions and should be kept confidential. Using the private key, the user can sign a transaction and can use receiver's wallet address to deliver. These addresses can be acquired by computing the cryptographic hash of sender's public key. The digitally signed transactions can be broadcasted throughout the network. There are two phases in a digital signature. (i) Signing phase, in which user A encrypts his data using his private key and sends user B the encrypted and original data. (ii) Verification phase, in which user B verifies the data using user A's public key. Through this process, it is possible to check whether the data is tampered or not. Digital signatures are generated using Elliptic Curve Digital Signature Algorithm (ECDSA) [9]. The Bitcoins ownership is determined by digital keys, digital signatures, and Bitcoin addresses.

1.8.2 KEYS

Bitcoins ownership rights are gained by keys, Bitcoin addresses, and digital signatures. The network does not store the digital keys. They're generated and stored in a file or database called "wallet." The keys inside a wallet can be issued and managed independently of the Bitcoin protocol by the wallet software. Keys can be generated even without the Internet. Keys allow many unique properties of Bitcoin, including decentralized trust, ownership certification, and encrypted – PoW model. Valid digital keys generate valid digital signatures which must be included in every Bitcoin transaction. Therefore, anyone who has a copy of the digital keys can acquire the Bitcoins in that account. Keys always come in pair, private and public keys. Like the bank account number, the public key can be known to anybody, but like the secret PIN number, only the owner can know the private key. Public and private keys are used to receive and sign transactions, respectively.

Bitcoin uses public key cryptography to generate the digital key pair. The public–private relationship is such that the private key produces signatures on messages and can be checked against the public key without revealing the private key. Bitcoin uses elliptic curve cryptography (ECDSA) for secured transactions. To be precise, the transactions are signed using the National Institute of Standards and Technology (NIST)-standardized ECDSA. For example, consider the transcription script of the standard "Pay-to-PubKey-Hash"; in order to prove the ownership, the user has to use his private key to provide his public key and signature. The user uses a per-signature random value to generate a signature. This value should be confidential and should be different for every transaction. Otherwise, it'll risk the private key computation. It is possible to extract a user's private key just by having a partially bitwise equivalent random values. Thus, a highly random and unique per-signature value for each transaction is necessary to keep the security of ECDSA. The relationship between private key, public key, and Bitcoin address is shown in Figure 1.4.

1.8.3 WALLETS

Bitcoin wallet is a software program which stores the Bitcoins. It is used to send and receive Bitcoins. Wallet is more like a spreadsheet that keeps track of the user's balance. In Bitcoin transaction, every Bitcoin is associated with an address. The various types of Bitcoin wallets are Software, Hardware, Paper, Brain, and Online wallets. Software wallets are the most commonly used and need a locally running Bitcoin instance. A reference implementation of the Bitcoin protocol which is a full client and processes the whole blockchain was released. Blochain.info or Coinbase are popular online wallets. Online wallets manage the wallets either centrally or using a hybrid approach where the wallet is encrypted and stored and the operations are performed at client-side browser. Software and online wallets are prone to security issues as the possibility of gaining access to the wallet is high if the intruder gains access to the target system. Hardware wallets solve the security problems as they use a separate device that works offline. As it is not connected to the Internet, it is hard for the intruder to gain access. Brain wallet is an advanced approach where keys are

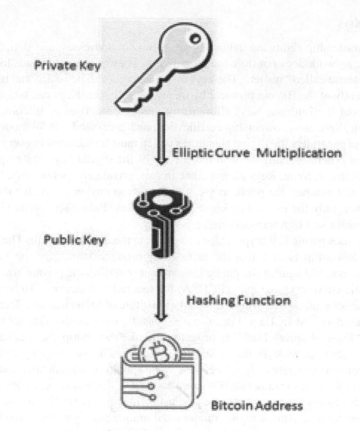

Private Key

Elliptic Curve Multiplication

Public Key

Hashing Function

Bitcoin Address

FIGURE 1.4   Relationship between private key, public key, and Bitcoin address.

stored in users' mind and retrieved by remembering a passphrase. Later, from this passphrase, the public key and the Bitcoin address are derived using hashing and key-generating algorithms. The passphrase should be relatively long to avoid any brute force attacks. Ref. [10] suggests that the security of the wallets can be improved by having multisignature transactions. Online wallets provide the kind of security where multisignature transactions are offered.

1.8.4   ADDRESSES

Receiver's public key is the address to which Bitcoins are being sent. Using sender's private key, the digital signature of the user is encrypted and can be decrypted using the sender's public key. Through this process, identification of the sender and the address of the receiver are verified. It is not possible to send Bitcoins if someone loses their private key. It is also possible for someone to steal the Bitcoins if he/she gets access to someone's private key. Even if the victim knows the intruders public key, he cannot take back his Bitcoins as he don't know the intruder's private key and the transaction cannot be reversed. Thus, every transaction has sender's and receiver's names to identify in blockchain. Public key cryptography is used in the identification process. The previous transaction history is hashed with the public key of the receiver, and the digital signature of the sender is added.

1.9   SECURITY ATTACKS ON BITCOIN SYSTEM AND COUNTERMEASURES

Since Bitcoin operates in a decentralized model with an uncontrollable environment, fraudsters and hackers consider transaction frauds are simple to do. Some of the common attacks on a Bitcoin system are (i) double spending, (ii) wallet attacks, (iii) network attacks, and (iv) mining attacks.

1.9.1   MAJOR SECURITY ATTACKS

1.9.1.1   Double Spending

Double spending is a situation where a fraudulent user tries to spend the same set of coins in two different transactions simultaneously. For example, a malicious user creates a transact T at time t using a set of Bitcoins Bc to a vender V to purchase something. At the same time, the user creates and broadcasts another transactions T′ using the same coins Bc, and sends the coins to his wallet address. This situation is called "double spending," where V will accept the purchase and send the goods, but cannot redeem the coins. For the coins are transmitted to the user's account. This problem can be solved by enforcing the rule by the network miners who will validate and process the transactions to ensure that coins spent in previous transactions are not used as inputs for the subsequent transaction. Also, the PoW bases consensus and time-stamping helps in the orderly storage of transactions in blockchain. In this case, when a miner receives T and T′ transactions, it can check if both transactions seek to use the same coins, and thus process only one transaction and reject the other. Figure 1.5 shows the double spending attack.

To detect the possibilities of double spending in payment systems, three techniques are suggested. The techniques are (i) listening period: in this, each transaction is linked by the seller to a listening duration, and he/she only delivers the product if he/she doesn't find any double spending attempt. (ii) Inserting observers: this is an extension of the first attempt in which the vendor employs few observers (nodes) under his control within the network. These observers inspect for any double spending and then send the transactions to the vendor. (iii) Forwarding double spending attacks: in this technique, the nodes forward the transactions which attempt to double spending to neighbors rather than discarding them.

1.9.1.2   Mining Pool Attacks

In order to reduce the verification time of a block, mining pools are created to increase the computing resources. This reduces the verification time, thereby winning a reward. Mining pools have a pool manager who sends the unsolved works to the pool members (miners). The miners derive partial proof of works (PPoW) and full proof of works (FPoW), and submit them as shares to the manager. If the miner

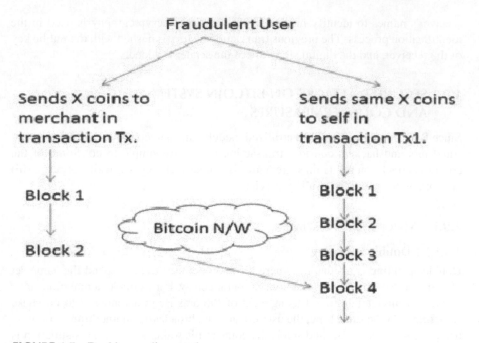

FIGURE 1.5   Double spending attack.

validates a block, it is then submitted to the manager. The manager then broadcasts the block over the network to get the reward. The reward will be distributed among the miners in pool based on the PPoWs. The pool manager decides the amount of work done by individual pool members based on the number of shares a member finds and submits while trying to find a new block. A share can be the solution of the mathematical puzzle. Mining pools are also subject to attacks internally as well as externally. Internal attacks are those where a set of malicious miners within the pools trying to collect more than their share or it may disturb the other miners to do successful mining acts. External attacks are those where the miners perform attacks using their hash power to do double spending.

The attacker in this case aims to disrupt the mining pool, rather than mining genuinely. Courtois and Bahack [11] utilize a gaming approach in which the miners use a passive mining strategy called "selfish mining" or "block discarding attack". Selfish mining is an attack in which the attacker hides the mined block initially and then reveals the block in such a way that it receives the share which is much bigger than their share of computing power. Or, it obscures the other miners and makes them waste their resources.

The most honest solution for pool attacks is suggested in Ref. [12]. In this approach, when the miner finds the presence of multiple forks in the same length, it passes the information to all other nodes and chooses one fork to extend. Similarly, other nodes also select some random forks and extend them so that the possibility of selfish mining can be minimized. Another solution is called "Freshness Preferred" (FP), where the blocks are time-stamped and the one with the recent timestamp would be proffered. The timestamp in this approach is unforgeable. To prevent the miners from using timestamps from the future, this approach uses Random Beacons [13]. Since block withholding is possible in selfish mining, this approach minimizes the reward as fresh blocks are preferred by the network.

1.9.1.3 Client-Side Security Threat

Because of its immense success, Bitcoin is attracting a growing number of users to join the network. Every Bitcoin client has a pair of private and public keys. Hence, there is a need to have secure key management techniques. Otherwise, if the keys are lost, the user will suffer monetary loss at a higher level. Since the keys are stored in a wallet, wallet thefts are high in nature using mechanisms like system hacking, buggy software installation, and wrong usage of wallet. The usage of public key cryptography raises the concern for secure storage and management of user keys.

In order to protect the wallet, "Cold Wallet" is proposed. It is a manual method and uses another account of the user where the excess amount of the user is stored. This method uses two computers in which one is not connected to the Internet, and a new private key is generated using wallet software. Since the second system is not connected to the Internet, it is not possible for the attackers to hack the wallet.

1.9.1.4 Bitcoin Network Attacks

Network attacks are majorly on Bitcoin and P2P communication networking protocols. The most common network attack is Denial-of-Services (DoS) that targets Bitcoin exchanges, wallets, and mining pools. As a Bitcoin network is distributed in nature, the effect of DoS is minimal on network operations. Hence, to disrupt the entire network, a strong attack like DDoS has to be applied. Unlike the DoS attack where a single attacker attacks the network, in DDoS, multiple attackers attack the network simultaneously. The intruder performs DDoS attack by removing the genuine miners from the network, thereby increasing the number of malicious users in the network. Under this situation, an honest miner may be burdened with numerous transaction requests from malicious miners. After some time, the miner may discard the input requests even the ones from honest miners. A total of 142 DDoS attacks were performed on 40 Bitcoin services, and 7% of known operators were the targets of these attacks [17].

Several solutions have been proposed to avoid or mitigate the network attacks. For DDoS attacks, a game theoretic approach is used. This approach presumes that the smaller pools are weighted less than the larger pools. Therefore, a balance is dawning between players and pools with greater participants getting more rewards than the smaller pools. Ref. [14] suggests that Proof-of-Activity (PoA) can withstand the DDoS attack. In PoA, a crypt value is stored in every block header by the user who places the first transaction, and thus any subsequent storage of transactions in this block is done only through honest miners.

All of the mentioned major attacks, in turn, have several other types of attacks which are mentioned in Figure 1.6.

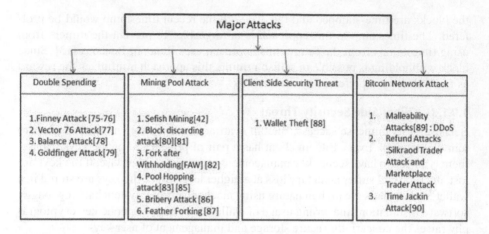

Major Attacks			
Double Spending	**Mining Pool Attack**	**Client Side Security Threat**	**Bitcoin Network Attack**
1.Finney Attack [75-76]	1. Sefish Mining[42]	1. Wallet Theft [88]	1. Malleability Attacks[89] : DDoS
2. Vector 76 Attack[77]	2. Block discarding attack[80][81]		2. Refund Attacks :Silkraod Trader Attack and Marketplace Trader Attack
3. Balance Attack[78]	3. Fork after Withholding[FAW] [82]		3. Time Jackin Attack[90]
4. Goldfinger Attack[79]	4. Pool Hopping attack[83] [85]		
	5. Bribery Attack [86]		
	6. Feather Forking [87]		

FIGURE 1.6   Security attacks in Bitcoin.

1.9.2   MINOR ATTACKS

1.9.2.1   Sybil Attack

In a Sybil attack, dummy nodes are installed by the attacker to make part of the network compromised. These compromised nodes perform a collective attack on the network. Sometimes the attacker hides or changes his identity and performs collision attacks using the compromised nodes. In collision attack, the attacker separates an honest miner and detaches the transactions initiated by the miner or manipulates the miner to choose the blocks that are supported by the attacker. If no miner approves a transaction, then it may be used for double spending attack.

1.9.2.2   Eclipse Attack

In a eclipse attack, the attacker creates a partition between the victim node and the public network by redirecting the victim's IP address toward the attacker. Figure 1.7 shows the network partition.

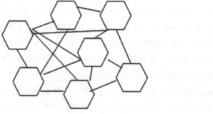

Network of Nodes

Isolated Nodes

FIGURE 1.7   Network partitioning.

1.9.2.3 Tampering

In a Bitcoin network, once a block is mined, the nodes broadcast the information to other nodes. This process happens periodically in the network. The network assumes that the information is broadcasted at a fair speed, but the attacker may delay the broadcast by creating some congestion in the network. This kind of tampering may lead to any of the network attacks discussed above.

1.10 PRIVACY AND ANONYMITY IN BITCOIN

In traditional models, a trusted third party shares and secures fewer details of the users in transaction. However, the trusted party knows all customer information. According to Bradbury [15], privacy means the context is hidden, and anonymity hides its owner. There are several instances where privacy is highly regarded than anonymity and vice versa. The concept of being anonymous is being unidentifiable and untraceable. It is difficult to ensure 100% anonymity in traditional models. Many applications such as Mixing services, Onion Routing, and The Onion Router (TOR) claim to be anonymous but happen to disclose identity information. Anonymity and privacy always come with a price. Either the system that claims to provide these two requires more resources in time, space, and computation or the user has to pay more for being anonymous and private. In Bitcoin, everything is transparent and all transactions are shared publicly. Hence, anonymity is provided by keeping the public key anonymous, i.e., using pseudonyms for the addresses. Since pseudonyms are used, it is assumed that Bitcoin is anonymous.

Nonetheless, it is widely stated that Bitcoin is not anonymous but described as "probably the world's most open payment network." All transactions are held in public, all users use pseudonyms of their addresses, and everybody in the network can track Bitcoin transactions between users. The real identity of users is not known as no other information of the users is stored. Even if there is a need for revealing identities between users, it will not be informed to any other users in the network. The users of Bitcoin cannot remain anonymous entirely, but they can be pseudonymous. It is believed that a certain level of privacy can be determined by the user while using Bitcoins. Using a new pair of keys (address) for every transaction can increase security to a large extent. Because creating a new key pair and linking it to every transaction avoids using keys that are linked to the previous transactions of the user, and thus the number of Bitcoins the user owns cannot be known. Another suggestion is, using multiple wallets for multiple purposes which cannot be linked to different transactions. For users who use wallets, their wallet transaction and address are tracked by the wallet service hosting it. So, it is possible for the hosting service to track the user's Bitcoin address through their IP address and disclose the user's identity. To prevent this, TOR, a popular tool, is used to hide the real IP address. TOR is a distributed overlay network consisting of TOR nodes for transmission control protocol (TCP)-based applications. TOR is designed based on Chaum's mixnets. In TOR, data encryption happens multiple times and is transmitted through the path over the nodes where decryption of each layer happens at each node.

Another Onion routing tool "I2P" (The Invisible Internet Project) [16] creates a hidden network within the network called darknet. I2P's structure allows multiple messages to encrypt inside the layers of encryption. The major drawback of I2P is the number of genuine users used to access the Internet through I2P is less. Few Bitcoin clients were developed to allow running and exchanging of Bitcoins with I2P. Other than TOR, Anoncoin is another coin intended to be fully I2P darknet coin.

After getting inspired by TOR, another new anonymization technology transaction remote release (TRR) was released. TRR was modeled to overcome the attacks that occur while using Bitcoin over TOR. TOR and TRR are similar in encrypting and routing the data, whereas transmission of Bitcoin transaction varies. Since TRR is specifically designed for Bitcoin, it encrypts and transmits only new transactions. However, TOR encrypts all blockchain data. This results in better productivity and performance of nodes. The only drawback of TRR is the need to change Bitcoin protocol. Although TRR is more vulnerable to DDoS attacks, it can withstand man-in-the-middle attacks.

1.11   RECLAIMING DISK SPACE

In order to save the disk space, it is possible to discard the spent transactions that happened earlier. To achieve this, Merkle tree hashes the transactions without changing the block's hash by adding only the root in block's hash. Through this, the need to store the internal hashes can be eliminated. Later, the old blocks can be compressed by cutting off the branches of the tree.

1.12   CONCLUSION

In this chapter, the P2P decentralized digital currency Bitcoin and its properties have been discussed. The discussion on cryptocurrencies before Bitcoin shows the state of the art of existing cryptocurrency. The transparency nature of Bitcoin is the key aspect to achieve validity and user anonymity is another aspect which lets Bitcoin to keep growing. The most significant concept of mining based on PoW secures the model and eventually stabilizes the consensus. The various security attacks are explained in detail along with countermeasures.

REFERENCES

1. D. Chaum, "Blind signatures for untraceable payments," in Proc. 2nd Conf. Adv. Cryptol., August 1982, pp. 199–203.
2. L. Lamport, R. Shostak, and M. Pease, "The Byzantine generals problem," *ACM Trans. Program. Lang. Syst.*, vol. 4, no. 3, pp. 382–401, 1982.
3. D. Malkhi and M. Reiter, "Byzantine quorum systems," *Distrib. Comput.*, vol. 11, no. 4, pp. 203–213, 1998.
4. J. Douceur, "The Sybil attack," in Proc. 1st Int. Workshop Peer Peer Syst., March 2002, pp. 251–260.
5. R. C. Merkle, "A digital signature based on a conventional encryption function," in Proc. 7th Conf. Adv. Cryptol. (CRYPTO'87), August 1987, pp. 369–378.

6. S. Haber and W. S. Stornetta, "How to time-stamp a digital document," *J. Cryptol.*, vol. 3, pp. 99–111, 1991.
7. R. C. Merkle, "A digital signature based on a conventional encryption function," in Carl Pomerance (ed.), *Advances in Cryptology—CRYPTO '87* (Lecture Notes in Computer Science), vol. 293. Heidelberg, Germany: Springer, 1988, pp. 369–378.
8. D. Eastlake, III and T. Hansen. U.S. secure hash algorithms (SHA and SHA-based HMAC and HKDF). 2011. [Online]. Available: http://www.ietf.org/rfc/rfc6234.txt.
9. V. S. Miller, "Use of elliptic curves in cryptography," in Hugh C. Williams (ed.) *Advances in Cryptology—CRYPTO'85* (Lecture Notes Computer Science), vol. 218. New York, NY: Springer-Verlag, 1986, pp. 417–426.
10. S. Goldfeder et al., Securing bitcoin wallets via a new DSA/ECDSA threshold signature scheme, Tech. Rep., 2015. [Online]. Available: http://www.cs.princeton.edu/stevenag/threshold_sigs.pdf.
11. N. T. Courtois and L. Bahack, "On subversive miner strategies and block withholding attack in Bitcoin digital currency," *CoRR*, vol. abs/1402.1718, 2014. [Online]. Available: https://arxiv.org/abs/1402.1718.
12. A. Sapirshtein, Y. Sompolinsky, and A. Zohar, "Optimal selfish mining strategies in Bitcoin," in Jens Grossklags and Bart Preneel (eds.), *Proc. 20th Int. Conf. Financ. Cryptography Data Security (FC)*. Oistins, Barbados: Springer, 2017, pp. 515–532.
13. M. O. Rabin, "Transaction protection by beacons," *J. Comput. Syst. Sci.*, vol. 27, no. 2, pp. 256–267, 1983.
14. I. Bentov, C. Lee, A. Mizrahi, and M. Rosenfeld, "Proof of activity: Extending Bitcoin's proof of work via proof of stake [extended abstract]y," *SIGMETRICS Perform. Eval. Rev.*, vol. 42, no. 3, pp. 34–37, 2014.
15. D. Bradbury, "Anonymity and privacy: A guide for the perplexed," *Netw. Security*, vol. 2014, no. 10, pp. 10–14, 2014.
16. F. Astolfi, J. Kroese, and J. V. Oorschot. I2P—The invisible internet project. Accessed: February 17, 2017. [Online]. Available: http://mediatechnology.leiden.edu/images/uploads/docs/wt2015_i2p.pdf.
17. M. Vasek, M. Thornton, and T. Moore. "Empirical analysis of denial-of-service attacks in the Bitcoin ecosystem." Financial Cryptography Workshops, 2014. Available: https://fc14.ifca.ai/bitcoin/papers/bitcoin14_submission_17.pdf.

2 Exploring the Bitcoin Network

Sathya A.R. and K. Varaprasada Rao
ICFAI Foundation for Higher Education
(Deemed to be University)

CONTENTS

2.1 INTRODUCTION

The most interesting feature of Bitcoin is its decentralized environment. Bitcoin is an unstructured peer-to-peer (P2P) decentralized digital currency. As Transmission Control Protocol (TCP) connections are the basic communication structure, Bitcoin is built on them. Bitcoin was invented by Satoshi Nakamoto in 2009 [1]. Transactions are carried out between Bitcoin users. All transactions over the Bitcoin network are recorded on a transaction log called "blockchain." The key feature of a currency is robustness and security. Cryptocurrencies attain these properties through a decentralized approach and cryptographic techniques. A decentralized method overcomes issues such as single-point failure and single source of trust, and increases the disagreement among different parties. Nevertheless, cryptocurrency uses a distributed mechanism that allows the device to maintain an unambiguous view of its state in order to reach consensus among users.

2.2   P2P NETWORK

A flat or hierarchical organization structure is used in P2P network, which arranges the peer in a random graph. Peers are interconnected over an unencrypted TCP channel. A peer or a node in a P2P network is a computer that has Bitcoin node software installed in it. This Bitcoin node software is added automatically in Bitcoin full client wallet. It is possible to view only the nodes which run full client, and not all the nodes can be seen while trying to locate a node in the network. If a peer is interested in entering the network, Domain Name System (DNS) server names and DNS seeds are first inquired. Such DNS seeds are hardcoded in Bitcoin clients to find the active nodes in the network. The DNS seeds return the peers' IP addresses that are ready to accept new connections in response to the enquiry completed. A version message with block version number and current time of the sender peer is sent to make a connection between peers. The receiver peer sends its version message back to the sender peer. Once the connection is established between the two, they acknowledge it by sending *verack* message. After joining the network, peer discovery takes place on the basis of address propagation mechanism. In the mechanism of address propagation, peers can ask each other's IP address using *getaddr* messages, and they can send their IP lists using *addr* messages. In order to find the peers that have some useful data, P2P uses techniques like expanding ring, random walks, Time-to-Live (TTL) search, etc. Peers can possess a sum of 125 connections, of which 8 are outgoing connections and 117 are incoming connections. Firewalls or network address translation (NAT) cannot have incoming connections and can have only outgoing connections.

Every peer maintains their list of connections made. When a new block is generated, the miner broadcasts the information to its peers. The receiver validates the block and then passes it on to his peers. When a peer makes a transaction, it sends the message to its related peers. A peer sends an *inv* message to initiate a transaction. If the recipient responds with a *getdata* message, then the transaction is sent using *tx* code. If the transaction is real, the transaction shall be communicated in the same manner as above. If any faulty messages are sent, the peer will be penalized by maintaining a penalty score. If the score reaches a threshold, the connection of the peer will be banned.

In general, the unstructured connections are easy to construct and strong in a dynamic environment, i.e., peers leave and join the network regularly. Searching a file in an unstructured network requires a flooding request as each peer would check on its local data. Due to a large number of copies of each question, there will be a huge overhead. Bitcoin network is, however, distinct from P2P file sharing networks. Bitcoin is not intended to find specific data objects to spread the data as quickly as possible and reach consensus over the network. P2P network structure is suited best for Bitcoin. Apart from this, the distributed model is used for data transmission, data storage, and data confirmation.

2.3   SUMMARY OF BITCOIN SYSTEM

The Bitcoin network is a worldwide decentralized consensus network built on a P2P architecture over the Internet. This network is formed across the world by individual nodes that run the Bitcoin core. Bitcoin core is an open-source software that enables consensus using a process called "mining." Through Bitcoin mining, the transactions are validated and recorded in an unchangeable, irreversible, distributed ledger called "blockchain." Bitcoin network allows transfer of digital money among the users.

Bitcoin is a digital money based on account entries. Bitcoin, therefore, cannot be perceived as digital currency but as balance of Bitcoin account. Bitcoin account is a key pair based on elliptic curve cryptography. The Bitcoin account can be located based on its address. The Bitcoin address can be generated using a public key. Any user can send Bitcoins based on the public information available to every other user. But to spend the Bitcoins, one needs a private key. A special software named "wallet" is used to create and manage this public–private key pair.

Payment happens between Bitcoin accounts through transactions. A Bitcoin transaction needs an input which is the source Bitcoin account's address and an output which is the destination address. A transaction can have one or more inputs and one or more outputs. From each input address, the amount of Bitcoins to be transferred is mentioned in transactions. Similarly, the amount to be transferred to each account from the destination address will also be available in the transaction. The source address is enforced to spend the exact amount received from the earlier transaction. Each input must, therefore, include the previous transaction identification number and the output index. As a result, at any point the output may either be spent or not spent. The output not expended is called Unspent Transaction Output (UTXO).

To authorize any transaction, the user of the input address must generate a digital signature using his private key. This process is to ensure the ownership of the account. The user who is at the receiving end has to (i) validate the digital signature for its correctness and (ii) validate the Bitcoins are not spent in any previous transactions before agreeing to the payment.

2.4 BITCOIN NODES

Although the nodes in the Bitcoin network are all identical, they can play various roles in the network based on the functionality they support. A Bitcoin node is a database consisting of a ledger, routing functions, mining, and wallet services. Figure 2.1 shows a full node with all functionalities.

A node that holds the whole blockchain up-to-date is called "full node," while few nodes only retain the blockchain subset and validate the transactions using a method called "simplified payment verification" or SPV. They are otherwise called as lightweight nodes. Mining nodes are aimed at creating new blocks to solve the proof-of-work (PoW) algorithm by having a special hardware. Some mining nodes are full nodes, while few are SPV nodes which are involved in pool mining. User wallets are nothing but Bitcoin clients and are part of full node. Wallets which are installed in resource-limited devices such as smart phones are considered as SPV nodes. Other than the main nodes, there are several other protocols run by the servers and nodes. These protocols can be mining pool protocols or lightweight client access protocols. The different types of nodes are mentioned in Table 2.1.

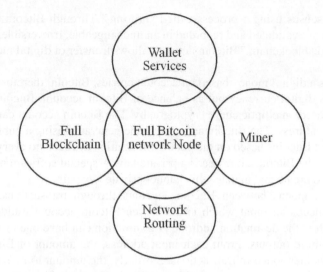

FIGURE 2.1   A Bitcoin network's full node.

TABLE 2.1
Bitcoin Nodes

Reference Client (Bitcoin Core)	Full Blockchain Node	Solo Miner	SPV (Lightweight) Wallet	Mining Nodes	SPV Stratum Wallet
On a P2P network, contains a wallet, full blockchain database, miner, and a network routing node.	On a P2P network, contains full blockchain database and a network routing node.	On a P2P network, contains a miner with full blockchain database and a network routing node.	On a P2P network, contains a wallet and a network routing node without a blockchain database.	On a P2P network, contains a miner without a blockchain and with other pool mining protocol nodes.	On a P2P network, contains a wallet and a network node without a blockchain on stratum protocol.

2.5   EXPERIMENTING WITH THE BITCOIN NETWORK

Connecting to the mainnet or the testnet is a good approach for learning about the Bitcoin network [2]. Testnet is the global space for experimenting with the Bitcoin protocols and their scripting potential. It applies a unique blockchain called "Faucets" and gives free coins. Except for few parameter alterations, testnet copies the mainnet and runs the same code as peers in Bitcoin. In contrast, more control is provided by local testing environment. Inbuilt with the Bitcoin reference software, a regression test (regtest) can generate blocks on request and can provide coins with no real value. This provides a safe environment to test new features. Bitcoin-testnet-box [3] is another similar approach. Both these approaches have been designed to use in circumstances where communications with random nodes and blocks are not preferred.

To simulate large-scale Bitcoin network, few simulation environments like simbit and Shadow are used. Both are discrete event simulators. Simbit literally exhibits the function of the Bitcoin client and a miner, and Shadow implements the Bitcoin software directly within the model system. It is identified that simulating exclusive blockchain functions and cryptographic operations prevents scalability of experiment. Therefore, only specific parts of Bitcoin client can be simulated so that both scalability and accuracy can be achieved as in real client behavior.

2.6 JOINING AND CONSERVING THE NETWORK

In the overlay, each peer in the Bitcoin network maintains a minimum of eight connections. Whenever this count is under run, the peer will actively try to make a connection with other peers. A peer can have more than eight connections if another peer tries to make a connection. The maximum number of connections a peer can handle is 125. The default port on which the peers receive incoming connection is 8333. To make a new connection, a peer implements an handshake by sending *version* and *verack* messages. Version message includes protocol version, timestamp for synchronization, and IP addresses. The peer node establishes a connection by responding with a *verack* message. The initial handshake between peers is shown in Figure 2.2.

If there are no messages for 90 minutes, the client will know the peer is offline. Bitcoin peers can also record the list of peers that are not connected directly to the network. After 24 hours in the network, every peer should send an *addr* message.

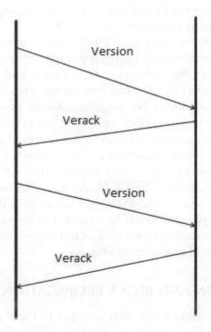

FIGURE 2.2 Handshake between peers.

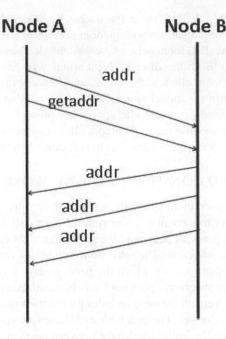

FIGURE 2.3 Exchange of an *addr* message.

If there is no message from a node, then the peer is considered as offline. The common way to explore the network is by exchanging the *addr* messages (during bootstrap). The *addr* message exchange is shown in Figure 2.3.

The peers must relay the address information for a maximum of ten addresses whenever an *addr* message is received. If a peer's timestamp information is no more than 10 minutes, the peer may attempt to relay the correct address. The address will be forwarded to one or two neighbors based on the peer's reachability in the network.

To find the neighbors during bootstrap, the Bitcoin peer uses three methods: (i) Internet Relay Chat (IRC), (ii) DNS, and (iii) asking neighbors. However, DNS is the default bootstrap mechanism since Bitcoin version 0.6. In 2014, to know information about other peers, *getaddr* messages are sent to initial peers. The same procedure was repeated for every unknown peer. Initially, 11,475 IP addresses were able to find, and after 37 rounds, 872,648 unique IP addresses were discovered. Studies have been able to locate network dimensions but not topology. In Ref. [4], AddressProbe, a technique to explore peer links, was developed. The *getaddr* message, given to a large network, helped to explore how the nodes learned from each other to map peers that were connected based on the timestamp.

2.7 TRANSACTION AND BLOCK PROPAGATION

In an unstructured network, the mechanism used to know information about the new transaction and block is simple, i.e., network will be filled with messages. Let us take an example of a transaction between peers and discuss in depth about the

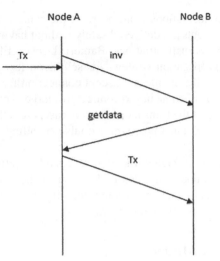

FIGURE 2.4   Transaction and Message Flow among nodes.

messages exchanged. The flow of messages between the peers is shown in Figure 2.4 below. Consider, user A has built a valid transaction. Prior to sending the transaction details like input and output, user A sends an inventory (*inv*) message to his neighboring peers, hinting that he is aware of new transactions. This inventory message contains a Transaction Ids list, but not the original transaction data. If the neighboring peers are unaware of any transaction, a separate *getdata* message will be sent requesting data about that transaction. Through this pull-based mechanism, excessive records of transaction can be minimized, thereby reducing the network load. After receiving *getdata* message, user A will send the corresponding transaction detail to the neighbor. Once the transaction is verified, it will broadcast the same way as user A did, initiating with *inv* message.

Bitcoin randomly selects one of every four messages to send *inv* messages and holds the remaining transactions. Thus, every neighbor receives random transactions. Although trickling reduces overhead, as in mix networks, it makes traffic analysis more difficult. It helps in hiding the creator of transactions. As messages are sent out for every 100 ms, there can be a delay in propagation of transactions. As Bitcoin trickling aims to transmit information as quickly as possible, it trades off overhead and piracy against the transaction of propagation. The user is responsible for his/her transaction. If a transaction does not get into a blockchain, then the user needs to rebroadcast the transaction and ensure that it gets added in the next block.

Three different types of relay patterns are detected from a highly connected peer. The most common relay pattern is (i) peer sending the transactions only once. The second relay pattern is (ii) receiving single or multiple transactions from the single peer. The third pattern includes (iii) multiple peers relaying a transaction and atleast one of them re-relays it. Knowledge of the transmission propagation helps to identify fraudulent users who pay with Bitcoin. For instance, consider a user wants to make a purchase using Bitcoins. The user wants to make a fast purchase without wasting time in waiting for confirmations from peer. At the same time, the seller wants to be assured that the products are delivered safely without having to worry about double spending. To address such situations, Bamert, Decker, Elsen, Wattenhofer, and Welten developed a propagation strategy for sellers on fast payments. As per [5], in such instances, the seller should not accept connection from the user as the only source of information. Rather, he has to connect to a large peer which, in turn, would force the user to broadcast the transaction. In this case, the seller should not relay the transactions. He would be able to recognize double spending as long as one genuine peer is connected to him.

Another approach is to signal the double spending attack by sending *alert messages* on a large number. The idea is that even if the opponent could trick the seller, some honest miners will immediately identify the crashing transactions and send warning messages over the network.

2.8 DEANONYMIZATION

Tracking message flows in a network helps to find the user information apart from understanding the network. Ref. [6] states that if the transaction is not facilitated by an online wallet, the IP address of a transaction can be determined by tracking the hub linked to all peers. This, however, disrupts the pseudonymous nature of the transactions. The relay patterns observed in Ref. [7] approve the expected behavior of the nodes. Based on the results from Ref. [6], the heuristics to match the transactions to IP address were developed even if the hub is not fully connected.

The Onion Router (TOR) [8], an Internet anonymity service, addresses this issue of trust by hiding the source node's IP. The Onion routing protocol is applied to a relay of network and thereby separates the sender and receiver information. Therefore, the source node's IP address is hidden, and the destination node can only display the address of the last TOR node. Bitcoin can handle the traffic through TOR using SOCKS interface. But there is a possibility of banning TOR connection in the Bitcoin network, which will lead to exploitation of DoS protection. Other than DoS attacks, it can also make the Bitcoin network susceptible to eclipse attack and man-in-the-middle attack.

Even in case of anonymized connections, it is possible to find the transaction's originator by observing the peer's neighbors serving fingerprints. As shown earlier, a client can make a minimum of eight connections and can broadcast the same to the neighbors in the network. These eight peers are called "entry nodes." Associations to these input nodes shall be maintained as long as they can be reached. This proves that the fingerprints are stable. An intruder can take advantage of this fact by generating a fingerprint for an IP address. This can be achieved by incorporating the fingerprint on all Bitcoin servers and recording the pairs forwarding the IP address. Thus, these peers are the entry nodes for the corresponding IP. Furthermore, only a fraction of entry nodes will be seen by an adversary due to the "trickling" of *addr* messages. This and the effects of timing can lead to false positives. Next, the intruder would be able to map the entry node transactions and thereby find the originator of the transaction. A solution proposed to mitigate this issue is to turn around the outbound connections. For example, the fingerprint can be blurred after every transaction. This gives rise to few entry selection policies for TOR anonymity network. Periodic entry node rotation will however create a high probability of selecting a malicious node. However, having a stable set of entry nodes will resolve the issue to an extent.

2.9 BOTNETS

Botnets are a group of computers connected to a computer that performs a repetitive task. Illegitimate botnets are installed in systems without the operator's knowledge to gain access to network resources and local files. Botnets can run any random program in computers. The botnets are linked to a bot master via the Command and Control (C&C) channel. The botmaster uses this channel to send and collect information from the bots. Blockchain can be used as a C&C infrastructure by programming instructions through transaction script [9]. The aim of botnet is to make through phishing and sending spams. It is also used for DoS attacks. In 2010, a study was conducted on Miner Botnet to understand the approach of bots in blockchain systems. The worker botnets would retrieve the graphic card information and initiates the mining software. The worker bots are connected to the proxy bots whose role is to run the Bitcoin client software and bind to a randomly chosen mining pool. This approach is called "proxied pool mining." There are approaches where bots are directly connected without proxies. The disadvantage of having a direct pool mining is, it is easy to detect them in the mining pools as they are large in number having the same account and has smaller hash rates. Another approach is the dark pooled mining in which the bots have their own mining pool and workers are connected to it.

2.10 BITCOIN-INSPIRED NETWORK APPLICATIONS

It is known that Bitcoin raised the digital currencies to a different level. There are also many applications inspired by Bitcoin, especially network applications; for example, decentralized domain name system [10], anonymous and distributed messaging [11], and abuse prevention of cloud services [12]. Naming services play an important role in computer networks. DNS, converting domain names to IP addresses, is a common application for the network. Zooko Wlcox-O'Hearn developed a concept called "Zooko's triangle." According to him, when a naming service is designed, it is possible to choose only two out of three properties: secure, distributed, and human-meaningful. For instance, OpenPGP public key fingerprints are distributed and safe, but are not human-understandable. Similarly, domain names are meaningful and safe, but are not distributed. According to Zooko, it is impossible to have a system with all these three properties incorporated. This phenomenon is called "Zooko's triangle."

Bitcoin-based naming services defy Zooko's triangle [13]. In Ref. [13], blockchain is assumed as key-value storage. Meaningful names were provided instead of assigning coins to the address. Like how Bitcoin stores transactions, the PoW scheme design of Bitcoin-based naming services protects mapping in blockchain.

Later, an alternative approach to DNS is given "namecoin." Namecoin shares Bitcoin's codebase and follows its characteristics. The mining process is similar to that of a Bitcoin and creates its own blockchain from a new genesis block. Bitmessage, a distributed, anonymous, and encrypted messaging protocol, is implemented. Unlike namecoin and Bitcoin, bitmessage does not store the messages forever; rather, it asks the peers to store the messages only for two days. To ensure the safe delivery of messages, an acknowledgment mechanism is adopted. The message broadcasting system is similar to that of Hashcash, where a PoW needs to be produced by the sender before broadcasting the messages. This bounds DoS attacks and spams.

2.11 BITCOIN NETWORK ATTACKS

Distributed Denial-of-Service (DDoS) is the most common networking attack. This attack mainly aims at the financial services of Bitcoin like eWallets, mining pools, Bitcoin currency exchanges, etc. It is difficult to implement DoS attack on Bitcoin network as it is distributed in nature. A single user launches the attack in DoS, whereas multiple users simultaneously launch the attack in DDoS. DDoS attacks are not much expensive but have the potential to disrupt the network at a higher level. In DDoS, the attacker aims at the honest miners and tries to take them out of the network so that malicious miners can be included in the network. It is observed that DDoS attacks on larger pools make a huge profit to the attacker than attacking individual or smaller mining pools.

Malleability attacks also aid DDoS attacks. The attacker can congest the transaction queue using *malleability attacks*. Transaction queue contains all pending transactions to be examined by the network. However, the attacker will create a fake transaction by keeping it as a high-incentive one. Thus, the miners are diverted to check this transaction and identify it as a fault transaction. By the time they realize, their time and effort must have been spent already. Malleability attack aims to waste the time and resources of peers in the network.

Bitcoin network is susceptible for routing attacks as well. Apostolaki et al. [14] discussed the effects of small- to large-scale routing attacks. Bitcoin's properties such as ease of routing and increased Bitcoin centralization with respect to routing and mining capability make it more vulnerable to the refund attacks. It is observed that the attacker who has taken control of <100 BGP prefixes can divide about half of the mining power.

Bitcoin payment protocol suffers weak refund policies which may lead to *refund attacks*. Such an attack on BIP70 payment protocol has been demonstrated successfully. BIP70 is the most commonly used payment protocol, and most of the wallets use this for Bitcoin exchange. The leading payment processors Coinbase and Bitpay use BIP70 for payment to more than 10,000 merchants. Refund attacks are two types: (i) Silkroad attack and (ii) Market-place trader attack. They deal with the weakness of existing payment policies in authenticating and manipulating.

Time jacking attack is another attack on Bitcoin networks. This attack manipulates the timestamps of the block by changing the network time with system time. As the creation of new blocks depends on network time counters, the attacker can deceive the node in accepting another blockchain if there are inaccurate timestamps. This attack may result in double spending, loss of miner resources, and reduced rate of verification of transactions.

The other attacks include *Sybil attack* and *Eclipse attack*. In Sybil attack, the attacker tries to control part of the network by including fake nodes. This is done by a group of nodes. In *Eclipse attack*, a victim node is separated from the public network by the attacker and IP addresses to which the victim nodes connected are routed to attacker's address. It is revealed that network congestion can be created by tampering the nodes of the network, thereby inducing several other attacks on network.

2.12   SECURING BITCOIN NETWORKS

To secure the Bitcoin network from the security threats discussed earlier, various countermeasures are presented in this section.

To review the *DDoS attacks*, Johnson et al. [15] gave a game-based approach where the mining pools compete with each other by increasing their computational costs. It was concluded through this approach that bigger pools will have more incentives than smaller ones. Eyal [16] also applied a similar gaming approach and identified that larger pools grab most of the incentives through selfish behavior in the network. Bentov et al. [17] suggested a Proof-of-Activity (PoA) protocol to resolve DDoS attacks. Another approach to solve DDoS attacks is to monitor the network traffic continuously using any user-defined web service or browser like TOR.

Time jacking is a deadly attack that has the potential to divide the network into multiple parts. To overcome time jack attacks, a set of techniques are suggested. These include the following: (i) to determine the upper limit of block timestamps, system time should be used instead of network time; (ii) possible time scales should be strictly followed; and (iii) only trusted peers should be used. To avoid *eclipse attacks*, the IP addresses of the trustworthy nodes are stored separately. Connections to the nodes are made on the trust factors of the nodes, and any mischievous nodes are banned from the network. It was advised that making few changes to the payment request message can reduce the probability of refund attacks. The payment request message should be modified with some customer information like registered email address, product information, or delivery address. Each payment request has a unique payment address associated with the key which would be used to make refunds. However, adding information about the customer may lead to compromise on user privacy. Anrychowicz et al. [18] provide a solution named "new deposit protocol" to deal with *malleability attacks*. It is proposed to save the transaction signature in a separate Merkle tree.

In general, transactions with huge amounts of Bitcoins are not carried out due to the fear of loss by malicious activities. In such cases, these transactions are broken into several smaller transactions, which ultimately delays the completion process as the network has to validate several transactions. Hence, to reduce this delay, payments through offline transactions, called "micropayments," are recommended. Micropayments take place via a separate channel called "micropayment channel." These channels are part of the Bitcoin network itself but committed means for counterparties to perform transactions.

Tampering attacks can be resolved when a peer broadcasts the time taken to mine a block along with the declaration of a new block. This would result in other peers in the network calculating the average time needed to mine a block. Through this, one can be assured that tampering of timestamp or unnecessary delay will not happen over the network.

In addition to the solutions discussed above, the use of Anti-Money Laundering (AML) regulations and Know Your Customer (KYC) policies with network traffic can also improve the quality of the mining process.

CONCLUSION

Data sharing can be achieved across all parts of the network simply by using an unstructured P2P network in Bitcoin. Security of Bitcoin networks is dependent on the efficiency of PoW-based consensus protocols which allow blockchain to maintain a consistent state. Blockchain inconsistencies could lead to effective double spending if properly exploited. Therefore, it is important that the Bitcoin network stays scalable in terms of size, storage, and bandwidth as it strengthens the consensus protocol. Full nodes are active in the P2P network and help to distribute data. Conversely, thin clients use an SPV to perform transactions. But as a fact, the network still has the general scalability problems of unstructured interfaces combined with the issues caused by the Bitcoin protocol itself.

REFERENCES

1. S. Nakamoto, "Bitcoin: A peer-to-peer electronic cash system," Technical Report, 2008 [Online]. Available: https://bitcoin.org/bitcoin.pdf.
2. Bitcoin. "Bitcoin developer documentation." 2014 [Online]. Available: https://bitcoin.org/en/developer-documentation.
3. github.com/freewil/bitcoin-testnetbox
4. A. Miller et al., "Discovering bitcoin's public topology and influential nodes," Technical Report, May 2015 [Online]. Available: https://cs.umd.edu/projects/coinscope/coinscope.pdf.
5. T. Bamert, C. Decker, L. Elsen, R. Wattenhofer, and S. Welten, "Have a snack, pay with bitcoins," in Proc. 13th IEEE Int. Conf. Peer Peer Comput. (P2P'13), September 2013, pp. 1–5.
6. D. Kaminsky. "Black OPS of TCP/IP, Black Hat USA." August 2011.[Online]. Available: http://dankaminsky.com/2011/08/05/bo2k11/.
7. P. Koshy, D. Koshy, and P. McDaniel, "An analysis of anonymity in bitcoin using P2P network traffic," in Proc. Financial Cryptography and Data Security: 18th Int. Conf., FC 2014, March 2014, pp. 469–485.
8. R. Dingledine, N. Mathewson, and P. Syverson, "Tor: The second-generation onion router," in Proc. 13th USENIX Secur. Symp. (USENIX Security'04), August 2004, pp. 303–320.
9. S. T. Ali, P. McCorry, P. H.-J. Lee, and F. Hao, "Zombiecoin: Powering next-generation botnets with bitcoin," in Proc. 2nd Workshop Bitcoin Res. (BITCOIN'15), January 2015, pp. 34–48.
10. Vinced. "Namecoin—A distributed naming system based on bitcoin." April 2011. [Online]. Available: https://bitcointalk.org/index.php?topic= 6017.0.
11. J. Warren, "Bitmessage: A peer-to-peer message authentication and delivery system," Technical Report, November 2012 [Online]. Available: https://bitmessage.org/bitmessage.pdf.

12. J. Szefer and R. B. Lee, "Bitdeposit: Deterring attacks and abuses of cloud computing services through economic measures," in Proc. 13th IEEE/ACM Int. Symp. Cluster Cloud Grid Comput. (CCGRID'13), May 2013, pp. 630–635.

13. A. Schwartz. "Squaring the triangle: Secure, decentralized, human-readable names," January 2011 [Online]. Available: http://www.aaronsw.com/weblog/squarezooko

14. M. Apostolaki, A. Zohar, and L. Vanbever, "Hijacking Bitcoin: Routing attacks on cryptocurrencies," in *Proc. IEEE Symp. Security Privacy (SP)*, San Jose, CA: IEEE, 2017, pp. 375–392.

15. B. Johnson, A. Laszka, J. Grossklags, M. Vasek, and T. Moore, "Gametheoretic analysis of DDoS attacks against Bitcoin mining pools," in *Proc. Financ. Cryptography Data Security FC Workshops BITCOIN WAHC*, Heidelberg, Germany: Springer, 2014, pp. 72–86.

16. I. Eyal, "The miner's dilemma," in *Proc. IEEE Symp. Security Privacy (SP)*, San Jose, CA: IEEE Comput. Soc., 2015, pp. 89–103.

17. I. Bentov, C. Lee, A. Mizrahi, and M. Rosenfeld, "Proof of activity: Extending Bitcoin's proof of work via proof of stake [extended abstract]y," *SIGMETRICS Perform. Eval. Rev.*, vol. 42, no. 3, pp. 34–37, December 2014.

18. M. Andrychowicz, S. Dziembowski, D. Malinowski, and Ł. Mazurek, "On the malleability of Bitcoin transactions," in *Proc. Financ. Cryptography Data Security FC Int. Workshops BITCOIN, WAHC, Wearable*, Heidelberg, Germany: Springer, 2015, pp. 1–18.

3 Blockchain Technology
The Trust-Free Systems

Sathya A.R.
ICFAI Foundation for Higher Education
(Deemed to be University)

Ajay Kumar Jena
KIIT (Deemed to be University)

CONTENTS

3.1 INTRODUCTION

Blockchain technology has revolutionized the world with its research work and business projects. Although blockchain sounds to be a new concept, it is there in the market for a while and only a couple of years ago it received the world's attention. In 2008, Satoshi Nakamoto explained blockchain as creation of a single or a group of people known by pseudonyms. Blockchain is a large database that stores the transaction information of people securely and allows them to interact with anyone without the need of trusting any third party. Technically, blockchain is a secured, shared, and distributed ledger that enables the process of recording and tracking resources without the need of a centralized trusted authority. It enables the sharing of information between two parties within a peer-to-peer network. The resources can be tangible like money, land, houses, car, etc. or intangible like digital documents, copyrights, intellectual property rights, etc. Blockchain is a distinct way of storing data using a "chained block" structure using cryptographic methods, thereby ensuring the integrity of the data.

3.2 HISTORY

Although blockchain has become popular recently, the distributed ledger technology (DLT) is a much older concept, which is often associated with blockchain and creates confusion between the two. Giaglis of the University of Nicosia says, "Both are methods of organizing transaction records in a shared, distributed database but, DLT is an umbrella term that encompasses all sorts of structures — including blockchain, which is just one type." The distributed ledgers need not to use blockchain, and blockchain does not have to be distributed though the practical value of blockchain is questionable without the characteristic of distributed ledger.

The concept of Blockchain had started in 1991 with Stuart Haber and W. Scott Stornetta's paper titled "How to Time-Stamp a Digital Document," which provided the details of procedures to be practiced while making or changing a digital document. This model used a cryptographically safe series of blocks to keep time-stamped documents. To keep many documents in a single block, Merkle tree was

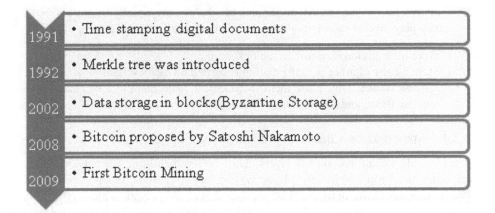

FIGURE 3.1 Timeline of blockchain technology.

introduced in 1992. However, this concept went unused much before Bitcoin's arrival. Later, David Mazières and Dennis Shasha in 2002 took the idea further and proposed how to store data using blocks. They focused on the multi-user network file system protocol and data structure, and proposed Secure Untrusted Data Repository (SUNDR) [1]. In 2004, Hal Finney, a cryptographic activist, proposed a system called Reusable Proof of Work (RPoW). The system takes a non-exchangeable Hashcash-based proof of work as an input and creates a Rivest, Shamir, Adleman (RSA)-based token that can be exchanged between users. RPoW keeps the ownership of the tokens in a trusted server. The users across the world can verify the correctness of tokens using this server. Through this, RPoW solves double spending issue in cryptocurrencies. This laid the framework for current blockchain technology. A major breakthrough in blockchain development happened in 2005 when a computer scientist, Nick Szabo, proposed a new blockchain-based currency called "Bitgold." This was one of the earliest attempts at decentralized currency. In 2008, the infamous anonymous pseudonym Satoshi Nakamoto invented peer-to-peer (P2P) mode electronic cash that allows direct transaction between two parties without having a trusted third party. The first block was mined in 2009, i.e., a transaction was checked and later added to the base of a network followed by a blockchain. Each block consists of verified transactions. Figure 3.1 shows the evolution of blockchain technology.

Starting from time-stamping of digital documents and hashing using Merkle tree through RPoW, blockchain has grown in a greater way.

3.3 OVERVIEW OF BLOCKCHAIN TECHNOLOGY

A blockchain is an open, transparent, distributed ledger that enables users to securely transfer unit ownership and can efficiently and permanently record transactions among users. Blockchain is a network of devices named "nodes" connected to each other over the Internet. Any electronic device connected to the Internet and having an IP address can be a node. All nodes on a blockchain are equally important, but a node may play diverse roles in making a blockchain work. A node can:

 i. Store information recorded on a blockchain.
 ii. Store a copy of all of the information recorded on a blockchain.
 iii. Process transactions, place them in blocks, append them to a blockchain, approve them, and forward them to the network.

3.3.1 STRUCTURE OF A BLOCK

A typical blockchain has six key layers. They are (i) data layer, (ii) network layer, (iii) consensus layer, (iv) incentive layer, (v) contract layer, and (vi) application layer. The bottom most layer in the architecture is data layer. It consists of blocks that hold a list of all transaction records like conventional public ledger. The transaction data can be a token exchange or any kind of information exchange over the network. Each block is split into two sections: the header and the body. Figure 3.2 displays a block's basic framework. All transactions are kept in the block body, while the block header retains the previous (parent) block hash. A block has only one parent. The first block of a blockchain is called the "genesis block," which has no parent block. By using a cryptographic hash, a block is recognized and linked to every other block so that the contents of it cannot be modified. If an attacker attempts to change the content of a block, their resulting hash value will also change the hash values of subsequent blocks. Therefore, if any hacker wants to alter a block, he has to change the hash values of all blocks in the series, which is practically not possible.

 i. Block version: defines what set of rules to follow for block validation.
 ii. Merkle root hash tree: hash values of all transactions of the block [2].

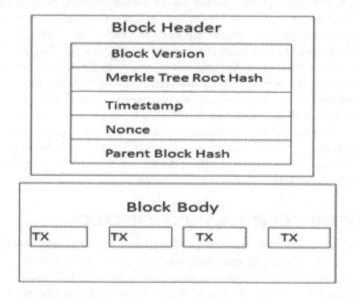

FIGURE 3.2   Basic structure of a block.

iii. Timestamp: current universal time in seconds as of 1 January 1970.
iv. nBits: target threshold of a valid hash block.
v. Nonce: A 4-byte field, usually starting at 0 and rising for each hash estimation.
vi. Parent block hash: a hash value of 256 bits that points to the previous block.

The block body is composed of a counter and transactions. The block size and the size of each transaction determine the total number of transactions a block can hold. The authentication of transactions is validated using an asymmetric cryptography method [3]. A fork will be created in the blockchain, if multiple nodes of the blockchain network produce valid blocks simultaneously. Under such a circumstance, the network chooses the longest chain as original and discards all the blocks in the other fork. A high-level view of blockchain is presented in Figure 3.3.

Blockchain is a P2P model where all peers are equally important. The transactions can be made between all nodes whenever they are generated. The nodes validate the transactions based on already-defined specifications. The transactions are forwarded only if they are valid. In this way, every node in the blockchain will have only valid transactions in its local copy. The validation of transaction authentication is based on digital signatures [4].

The layer above network layer is the consensus layer. In a decentralized environment and among non-trustworthy nodes, the process of achieving consensus is handled by the consensus layer. The consensus layer is comprised of different consensus algorithms. The four major consensus algorithms are Proof of Work, Proof of Stake, Delegated Proof of Stake (DPoS), and Practical Byzantine Fault Tolerance (PBFT). Apart from these algorithms, there are few less popular consensus algorithms such as Stellar, Proof of Bandwidth (PoB), Tendermint, Ripple, Proof of Authority (PoA), Proof of Burn, Proof of Elapsed Time (PoET), Proof of Space, Proof of Retrievability, Proof of Activity, Proof of Trust, BFT-SMART, Proof of Luck, and Scalable BFT.

The incentive layer takes care of the economic factors of the blockchain network. It encourages the nodes to provide their contribution for verifying the data by integrating the incentives allocation mechanism and incentive issuance process. Thus, the incentive layer acts as a motivating force of the blockchain network.

The programming part of the blockchain is handled by the contract layer. Smart contracts, various algorithms, and scripts are all dealt in this layer. A smart contract

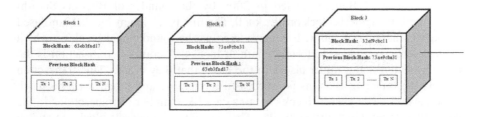

FIGURE 3.3 Blockchain structure.

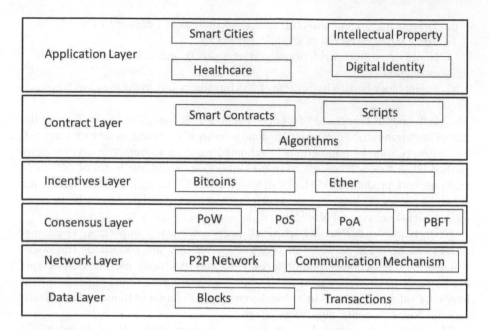

FIGURE 3.4 Blockchain breakdown structure.

is a collection of self-executing procedures recorded in a blockchain. Using smart contracts, the digital assets can be controlled, business logic can be explained, and user rights can be formulated. Once the parties involved in the transaction agree to the conditions mentioned in a smart contract, it will be cryptographically signed and sent over the network for verification. The smart contract will automatically be executed as per the mentioned rules if the predefined conditions are met. The well-known programming language used to write smart contracts in Ethereum is Solidity [5].

The highest layer in the architecture is the application layer. It is comprised of business applications like Internet of Things, healthcare, digital identity, market security, etc. Different layers of blockchain architecture are presented in Figure 3.4.

3.3.2 THE GENESIS BLOCK

The first block of Bitcoin created in 2009 by the founder of Bitcoins Satoshi Nakamoto is the genesis block or Block # 0. The first block in any blockchain-based protocol is the genesis block. It is considered as the foundational block on which further blocks are added to form a chain of blocks. A coinbase transaction is the first transaction that a miner places in the block that they create. This transaction rewards the miner in Bitcoins for successfully creating and broadcasting a block in the network. The genesis block contains data like number of transactions in the block, timestamp, block difficulty, etc. The elements of a genesis block are shown in Figure 3.5.

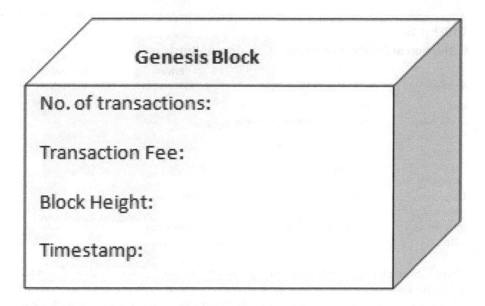

FIGURE 3.5 A genesis block.

3.4 TYPES OF BLOCKCHAIN

Based on the application of the blockchain, it is classified into three major types as below. Each type has some unique features than the other.

3.4.1 Public Blockchain

Public blockchain is an open-source and rightly decentralized setup in which anyone can be users, miners, or developers, and anybody can participate and publish new blocks. Thus, all transactions over public blockchain are transparent. They are also called "permissionless" as it permits anyone to take a copy of the blockchain and involve in block validation. Ethereum and Bitcoin are the best examples of public blockchain. As public blockchain allows a large number of users to be anonymous in network, it is mandatory to minimize the risk of malicious behavior. Therefore, publishing a new block involves solving a computationally difficult puzzle. A transaction fee is included in each transaction and is given as a reward to nodes trying to publish a block.

3.4.2 Private Blockchain

Private blockchain is a permissioned blockchain, which means participants need consent to join the network. Only users who are given the permission to join the network can view the transactions. Since entities running the chain have more control over the participants and governing structures, private blockchain is more centralized than public blockchain. They are more suitable for individual enterprise solutions.

TABLE 3.1

Evaluation of Public, Private, and Consortium Blockchains

Attribute	Public Blockchain	Private Blockchain	Consortium Blockchain
User access	Public	Restricted	Restricted
User identity	Anonymous/pseudo-anonymous	Approved users	Approved users
Participation in consensus	All peers	Single enterprise	Selective peers in multiple enterprise
Content immutable	Yes, fully	Partly	Partly
Transaction processing rate	Low	High	High
Permissionless	Yes, Permissionless	Permissioned	Permissioned

There is no exchange of tokens or currency in private blockchain. No transaction fee is involved as nodes validating the block are known to each other. Hence, private blockchain is not tamper resistant like public blockchains. It is possible to rollback in private blockchain if the enterprise wishes to at any point of time. Hyperledger is an example of private blockchain.

3.4.3   Consortium Blockchain

Similar to private blockchain, consortium blockchain also needs permission. On the other hand, the network expands to several organizations and provides account-ability between the parties involved. Transaction fee is not involved in consortium blockchain, and computationally it is not difficult to publish a block. Consortium blockchain has the privacy benefits of private blockchain and has the secure and transparent natures of public blockchain. It is also called "hybrid blockchain." Dragonchain is an example of hybrid blockchain. Table 3.1 summarizes the features of different blockchain types.

3.5   PROPERTIES OF BLOCKCHAIN

Many researchers work on blockchain technology, and their findings are influential in developing the technology suitable for many applications. Following are the salient features of blockchain technology.

3.5.1   Decentralized

In traditional centralized transaction systems, transactions are made through a mediator that provides transaction services. Usually, banks act as the central authority of the system that controls everything. The main aim of the decentralized system is to remove the necessity of a central authority and create a decentralized system across the world with computers (often called "nodes") as parts of the system. Thus, all the nodes in the network will have the control instead of having one single authority.

3.5.2 Transparent

All transactions that take place on blockchain are open, and thus, anyone who is a part of the network can view all the transactions. This is the transparency that blockchain provides which never existed in the centralized system. Although blockchain provides anonymity of the users, it is possible to view the transactions made by the users using their public address.

3.5.3 Immutable

Immutable means once the data is entered in blockchain, then no changes can ever be made in that data. This property provides security and assures the users that the data will not be altered.

3.5.4 Persistency

Validation of transactions can be very quick, and invalid transactions can be identified and never are allowed in the block. Once the transactions are connected to a block, it is not possible to remove or reverse them.

3.5.5 Anonymity

Any user can communicate with every other user in the network using their public address. It is not possible to identify the user using this address, and anonymity is maintained without compromising the system's transparency, as all transactions are documented in a public ledger. Yet, many researches show, using the right clustering and flow analysis technologies it is possible to track the user's identity. Nonetheless, a lot of work has been done to address the privacy and anonymity issues, and various schemes are proposed to improve anonymity property.

3.5.6 Auditability

Blockchain transaction stores customer balance data based on an Unspent Transaction Output (UTXO) model. Whenever a transaction takes place, the unspent transactions will be referred. Once the transaction is complete and documented into a block, the status of the unspent transactions will be changed to spend. Hence, the transaction can be easily verified and tracked.

3.6 TRANSACTIONS AND DIGITAL SIGNATURE

A transaction or any data exchange to take place in a blockchain, the nodes of the network need a private and public key pair. A node uses a private key to sign and send the transaction to the receiver node's blockchain address. These addresses can be obtained by cryptographic computations of the receiver's public key. SHA-256 encryption algorithm is used to calculate the blockchain address. To understand the transaction over blockchain, let's take an example of a bus company. We book tickets via an app or web. The credit card company charges to process the transaction.

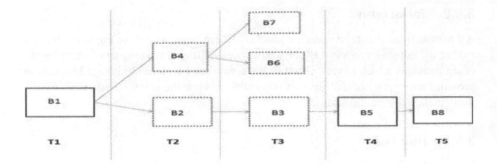

FIGURE 3.6 Blockchain fork.

Using blockchain, the bus operator can not only save on credit card processing fees but can also move the entire ticketing process to the blockchain. The parties involved in the transaction are the passenger and the bus company. The ticket is a block, which will be added to a ticket blockchain. This transaction on a blockchain is a unique, independently verifiable, and trustable record. The ticket blockchain is also a database of all transactions. For example, a certain train route, or the whole train network, that includes every journey ever taken, every ticket ever sold. The entire transaction on blockchain is free. The blockchain not only transfers and stores money, but it can also replace all processes and business models which charge transaction fees.

Transactions are not necessarily to happen within single blockchain. It can happen between two separate blockchains as well. This is called sidechaining [6]. Sidechains are also blockchains that run parallel to the main chain (existing blockchain) and are synchronized. It is possible to exchange tokens from the main chain to the sidechain and vice versa.

The longest blockchain is considered as the real and latest one. As the block validation process is distributed in nature, there could be a possibility of having two valid solutions at the same time. This results in creation of forks in blockchain. But, blockchain is an ordered set of transactions, and the miners need to maintain a consistent view of it. The miners, however, are free to select a fork and continue to mine. But once the miner working on a fork announces a valid block, a long blockchain is now shown to the network and users continue linking their blocks to it. Figure 3.6 shows the forks happened over the mining process.

3.7 MERKLE TREE

Merkle tree is otherwise termed as Binary Hash Tree [7]. Generally, it is used to process the huge amount of data. In blockchain, Merkle tree is used to organize the transactions in such a way that it requires only fewer resources to process. Let us see an example using Merkle tree shown in Figure 3.7.

After each transaction is hashed separately, they combine their hash values with the nearby transaction and are hashed again together. In Figure 3.7, even-numbered transactions are shown; if there are odd-numbered transactions available for hashing, then the last hash is simply combined with itself and is hashed to produce a new

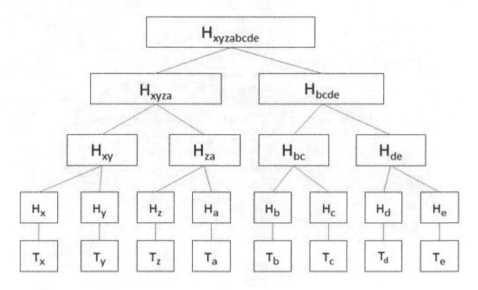

FIGURE 3.7   Merkle tree.

hash value. This process of combining and hashing will be repeated until the last hash value is obtained. This value is termed as "Merkle root." Here in this example, $H_{xyzabcde}$ is the Merkle root which is the hashed value of all the transactions. The size of Merkle root is 32 bytes and will be placed inside the block header. Thus, the Merkle root inside the block header represents the summary of all transaction data. If there is any change in any of the transactions, then the corresponding hash value will also change, resulting in a change in combined hash value and finally a different Merkle root. Hence, the Merkle roots help in finding whether any transaction in a block is tampered. Using Merkle root, it is also easy to check if a specific transaction is added to the block without having to download the entire blockchain. For example, if the user needs to check whether T_a is available in the block, the only value the user needs is the Merkle roots of H_{xy}, H_z, and H_{bcde}.

3.8   SHA-256

SHA-256 is a cryptographic hash function which takes an input of any size and produces a fixed-size output [8]. Hash functions are more powerful because they are unidirectional, i.e., if an input is given, an output can be generated by anyone. Nevertheless, the input value can't be recreated using the hashed output value. The powerful nature of SHA-256 makes it perfect for blockchain applications.

3.9   HOW BLOCKCHAIN WORKS

Blockchain is a distributed, decentralized public ledger that records transactions across the network (see Figure 3.8). The way in which blockchain works can be explained with an example. Let's say, User A needs to send money to User B. To make this monetary transaction, A needs to know B's wallet address. A wallet

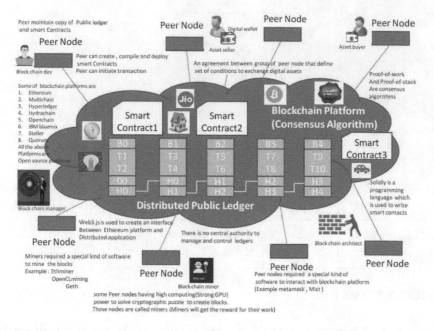

FIGURE 3.8 Blockchain transaction.

address is more like an email address, to which we send money instead of email. The transaction is initiated once A sends the money to B. In order to get this money in B's wallet, every node in the blockchain network has to verify the transaction. During this process, every node will record the transaction. Hence, thousands of records will be stored in thousands of computers. Users may not like the idea of having their transaction information available on many computers. Blockchain resolves this by keeping the transaction anonymous. To maintain the integrity of transactions over the network, the nodes use a consensus protocol, cryptographic hashes, and digital signatures. Consensus protocol assures that the public ledgers are exact copies and minimizes the number of fraudulent transactions. If any of the users wants to change any transaction record, they need to have access to all the nodes in the network which is practically not possible. SHA-256, a cryptographic hashing algorithm, is used to assure that any change in a transaction leads to a new hash value being computed that demotes a transaction input that may be compromised. Digital signatures are used to assure that the transactions are initiated from genuine senders and not from any imposters. "Miners" are people dedicated to devices and blockchain. Devices are the nodes on the network which validates transactions and stores them in a distributed ledger. These miners are rewarded in Bitcoins for their services.

3.10 BLOCKCHAIN OPEN-SOURCE IMPLEMENTATION

Several open-source implementations of blockchain technology are available, and selecting one among them to use is a challenging task. A comparison of several blockchain implementations is made in Table 3.2.

TABLE 3.2

Comparison of Open-Source Blockchain Models

Properties	Bitcoin	Ethereum	Hyperledger	Corda
General description	Generic blockchain platform	Standard blockchain platform	Modular blockchain platform	Special distributed ledger for Fin. Tech.
Governing authority	Public	Bitcoin developers	Linux foundation	R3
Operation mode	Permissionless—public	Permissionless—private or public	Permissioned—private	Permissioned—private
Smart contract	No	Yes (code in Solidity)	Yes (code in Go, Java)	Yes (code in Java, Kotlin)
Consensus	PoW	PoW/PoS	PBFT	No specific mechanism
Currency	Bitcoin	Ether	None	None
Pros	Can accommodate any number of nodes in the network	Scalable and not much of computation required for PoS	Faster compared to all other frameworks	Privacy and interoperability
Cons	Time taking and Computationally expensive	Needs wealth or stake	Cannot scale more nodes	Cannot be used outside financial sector

3.10.1 BITCOIN

Bitcoin's design is public and not owned or controlled by anyone, and everyone can be a part of that network. Refer Chapter 1 for a detailed discussion on Bitcoin [9].

3.10.2 ETHEREUM

Ethereum is an open-source software based on blockchain technology [10]. It is a distributed public blockchain network. It enables the developers to develop and install decentralized and distributed applications. Ethereum is similar to Bitcoin, except that the purpose and capability of both differ. One specific application of blockchain, i.e., P2P electronic cash system, is offered by Bitcoin. In contrast, Ethereum emphasizes running the code of any decentralized applications. In Ethereum, miners work to get ether unlike Bitcoin network where the miners work to mine Bitcoins. Ether is another type of crypto token that runs the Ethereum network. Ethereum has another type of crypto token called gas which would be issued as transaction fee for the miners.

3.10.3 HYPERLEDGER

Hyperleger is another open-source blockchain framework developed by a group of cross-industry experts. It adopts a collaborative software development approach that assures security, transparency, and interoperability. Hyperledger offers a wide range of tools and frameworks to create enterprise-based solutions.

3.10.4 CORDA

Corda is another open-source blockchain platform used for collecting and processing financial agreements that abide smart contracts. Any service that is built on Corda is compatible with network users. Corda is the only platform that offers both privacy of private networks and interoperability of public networks.

Other than the discussed open-source platforms, there are quite a number of other popular frameworks such as Quorum, Openchain, Hydrachain, and Multichain.

3.11 TESTNET

An alternative Bitcoin blockchain used for testing is Testnet [11]. Put simply, Testnet is a replica of blockchain running in a host computer and primarily used for testing purposes. Testnet is not only for Bitcoin but also for every blockchain implementation at some point in the source code. This is because Testnets are vital components in the development process of a functional blockchain. Testnet coins do not have any monetary value and are different from actual Bitcoins. Since it has no financial value, experimenting is simple for the application developers or testers, without having to use expensive Bitcoin. Three generations of Testnets have existed thus far. Testnet2 was the very first Testnet reset to be done with a new genesis block since people were starting to exchange Testnet coins for legit money. The currently running test network is Testnet3. There are two major differences between Testnet and mainnet: (i) networkId—a network id is an identifier for the network and (ii) genesis block—the very first block of the chain. Since these two work on entirely different networks, they both have different genesis blocks though the content of the blocks may be the same.

As compared to the main blockchain, Testnet receives fewer transactions and is usually much smaller. Since January 2018, the disk storage size was 14 GB, containing data for Testnet operation estimated for about six years. To download this data, approximately 12 GB of network activity peaking at 2 MB/s transfer rate is required. Testnet performs few important functions as follows.

3.11.1 CONTINUOUS DEVELOPMENT

Blockchain technology is still in its infancy and needs huge amount of development and testing to achieve mainstream acceptance and practice. In order to overcome the various challenges of blockchain, a lot of research and rigorous testing is required. Ultimately, Testnet acts as a simulation on how a real chain would work under real-world conditions.

3.11.2 PREVENTS DISRUPTIONS

As Testnets are replica of real chains, it is evident that conducting tests on a real chain would be more difficult as a complex interaction among the components can disrupt the network and can even break the chain. Hence, Testnet helps to ensure that everything is functioning properly before launching real chains.

3.12   PAYMENT VERIFICATION

In Bitcoin network, simplified payment verification nodes will have a subset of blockchains, i.e., only block headers. Simplified payment validation (SPV) nodes largely use Merkle trees. To check whether a transaction is added in a block, it uses a Merkle path or authentication path. This eliminates the need to download all the transactions. Consider, for example, an SPV [12] node that is willing to receive payments to an address in its wallet. A bloom filter is mounted on its peer-related connections by an SPV node to restrict the transactions received to those with interest addresses only. A merkleblock message will be used to submit the block if a peer recognizes a transaction that matches the bloom filter. The merkleblock message contains a Merkle path that connects the transaction specific to the Merkle core of the block and a block header. The SPV node can use this Merkle path to connect the transaction to the block and can ensure that the transaction is included in the chain. An SPV node uses the block header to connect the block to the rest of the blockchain. The ties between block and transaction and the connection between block and blockchain indicate that the transaction is inserted in the blockchain.

3.13   BENEFITS AND CHALLENGES

3.13.1   BENEFITS

3.13.1.1   Decentralization
There is no third-party intervention during transactions. Whatever cryptocurrency the user owns, it belongs only to the users. No central authority can own or handle it. Decentralized network provides protection against corruption and tampering.

3.13.2   TRANSPARENT AND ANONYMOUS

It is not easy to trace down a user's address unless the user shares the wallet address. Even if the user publicizes, it is easy to generate another wallet address. Hence, the privacy of the user will be maintained. Simultaneously, as all blockchain transactions are registered in a public ledger, blockchain retains confidentiality without losing the transparent nature. While few research works have said that a transaction in blockchain can be traced back using clustering and flow analysis technologies, many research studies have suggested different schemes to boost anonymity property.

3.13.3   LESS TRANSACTION FEE AND NO TAXES

As blockchain transactions are decentralized and anonymous in nature, there is no practical way to apply any tax, and for instant transactions, there is no cost involved. Even if there is any cost involved, it is much less than a credit card, Paypal, or bank transfer [7].

3.13.4   THEFT RESISTANCE

It is not possible to steal Bitcoins from any user unless the rival knows the private key that is associated with the user's wallet. However, if the user loses the private key associated with his/her wallet by any means, the bitcoins in the wallet also lost forever and can never be recovered.

Though blockchain has numerous benefits for the betterment of the Internet, it does have to face some technical challenges as well.

3.13.5 SCALABILITY

Scalability is of big challenge. The size of Bitcoin is limited to 1 MB, and mining a block takes 10 minutes. The Bitcoin network is incapable of dealing high-frequency trading since it is limited to a rate of seven transactions per second. However, if the size of the block is increased, the storage space will also be increased and it will slow down the network speed. This results in centralization because fewer consumers want such large networks to be managed. The tradeoff between security and block size was therefore of major concern.

3.13.6 HIGH ENERGY CONSUMPTION

The network blockchain uses the PoW model to achieve consensus. Although the PoW model protects the mining process from various security threats such as Sybil [8] attacks and double spending, it requires a great deal of energy and computing resources to calculate the required hash value for a block.

3.13.7 SELFISH MINING

In blockchain, selfish mining strategy allows a miner to achieve larger revenue than the fair share. Miners may in future hide their mined blocks for more revenue. This can lead to many divisions of blockchain and hinder the growth of blockchain.

3.13.8 PRIVACY LEAKAGE

Privacy at certain level can be achieved through public and private key pairs. Users transact through this key pair without knowing the real identity. Transactional privacy is not guaranteed by blockchain since the transactions and balances for each public key are available in public. In addition to this, the identity of the user can also be monitored through his/her transactions.

CONCLUSION

Blockchain, in simple words, can be defined as a series of immutable, time-stamped data blocks managed by a set of computers not owned by any single entity. Data are encrypted in each of these blocks and clustered together using cryptographic principles. The blockchain network has no central authority – the very definition of a decentralized system. As it is a public and permanent ledger, the information contained in it is open for access by anyone and all. So, anything that is developed on the blockchain is transparent, and their actions are accountable to everyone concerned.

REFERENCES

1. D. Mazieres and D. Shasha, *Building secure file systems out of Byzantine storage*, ACM, New York, NY, 2002.
2. M. Pilkington, *Blockchain technology: Principles and applications*, Research Handbook on Digital Transformations, Edward Elgar Publishing, Cheltenham, UK, 2015.
3. A. Kosba, A. Miller, E. Shi, Z. Wen, and C. Papamanthou, "Hawk: The blockchain model of cryptography and privacy-preserving smart contracts," in IEEE Symposium on Security and Privacy, 2016, pp. 839–858.
4. METI. "Survey on blockchain technologies and related services," Technical Report, December 2017. [Online]. Available: https://www.meti.go.jp/english/press/2016/pdf/0531_01f.pdf.
5. Solidity. "Solidity," June 2018. [Online]. Available: https://solidity.readthedocs.io/en/develop/.
6. A. Back, M. Corallo, L. Dashjr, M. Friedenbach, G. Maxwell, A. Miller, A. Poelstra, J. Timón, and P. Wuille, "Enabling blockchain innovations with pegged sidechains," Technical Report, 2014. [Online]. Available: https://blockstream.com/sidechains.pdf.
7. R. C. Merkle, "A digital signature based on a conventional encryption function," in *Advances in cryptology—CRYPTO '87* (Lecture Notes in Computer Science), vol. 293. Springer, Heidelberg, Germany, 1988, pp. 369–378.
8. D. Eastlake III and T. Hansen. US secure hash algorithms (SHA and SHA-based HMAC and HKDF), RFC 6234 (informational), internet engineering task force, May 2011. [Online]. Available: http://www.ietf.org/rfc/rfc6234.txt.
9. Z. Zheng, S. Xie, H. Dai, X. Chen, and H. Wang, "An overview of blockchain technology: Architecture, consensus, and future trends," in Proc. IEEE BigDataCongress'17, Honolulu, HI, USA, June 2017, pp. 557–564.
10. G. Wood, "Ethereum: A secure decentralised generalised transaction ledger," *Ethereum Project Yellow Paper*, vol. 151, 2014, pp. 1–32.
11. https://en.bitcoin.it/wiki/Testnet
12. Z. Zheng, S. Xie, H. Dai, X. Chen, and H. Wang, "An overview of blockchain technology: Architecture, consensus, and future trends," in Proc. IEEE BigDataCongress'17, Honolulu, HI, USA, June 2017, pp. 557–564.

4 Consensus and Mining in a Nutshell

Sathya A.R.
ICFAI Foundation for Higher Education
(Deemed to be University)

Santosh Kumar Swain
KIIT (Deemed to be University)

CONTENTS

4.1 BACKGROUND

Bitcoin is a peer-to-peer digital currency used for Internet commercial transactions. Basically, Bitcoins have no value, and those who trade them decide the worth of it. Bitcoin interests many due to its user privacy nature, facts behind the transactions, its suggestion as a payment system, and many more. The peer-to-peer network participants maintain a distributed Bitcoin transaction ledger which is accepted by the network. Bitcoins are stored in this ledger based on Bitcoin addresses. The public keys from a key pair serve as Bitcoin addresses. In order to transact the Bitcoins to another user, the current user (owner) of the Bitcoin has to sign a transaction using the private key of the key pair by Elliptic Curve Digital Signature Algorithm (ECDSA). Once the transaction is validated and the signature is verified by the network's other nodes, the transaction will be accepted by the network and will be added to the ledger.

There is no central authority to issue or control the Bitcoins, but it is created by a process called *mining*. Mining is a process in which all legitimate transactions are stored in blocks and included in the register called "blockchain." The added blocks are linked to the previous blocks so that reusing of Bitcoins can be prohibited. To add a block to the blockchain, a peer requires a nonce value that satisfies a particular equation using SHA-256 cryptographic hash function. Finding a nonce value is a computationally expensive task. So, a node who finds the suitable nonce value will be rewarded with an incentive. Nevertheless, this method of verifying a transaction might be confusing when each node attempts to relay its found block. Hence, to avoid such situation, a consensus algorithm is made between all peers in the network to decide which blocks should be added, which peers should be allowed to include the block, and so on.

Consensus algorithms aim to safely update repeated distributed states and are the key piece of the puzzle in the operating principles of the blockchain. Consensus protocol assures that copies of the shared state are synchronized and accepted in blockchain at any given time. According to Refs. [1] and [2], the deterministic consensus in entirely asynchronous models of communication cannot recognize any faults. Therefore, partial synchronization assumptions are required with maximum transaction propagation latency thresholds. Cryptography and partial synchrony, digital currency and precursor designs are some of the previous works on consensus protocols. These were the building blocks of blockchain networks to create "decentralized" consensus algorithms.

Several consensus algorithms have been proposed till date, which will be discussed in later sections of this chapter.

In 2009, many variations of Bitcoin, such as Ethereum [3] and Nxtcoin [4], were introduced. A variant in which anyone can join and leave the network anytime is called "public blockchain." Since the nodes can join and leave anytime, it is hard to predict the exact number of nodes; hence, applying the concept of consensus is difficult. In case the number of nodes is too high, exchanging the agreement and taking consensus of all nodes is even more difficult. On the other hand, to add a block in a blockchain, the nodes have to prove their quality compared with other nodes. Hence, a consensus algorithm named "proof-based consensus" was proposed.

Proof of work (PoW) [5] is the first version of a proof-based algorithm. In this consensus algorithm, a node can broadcast their blocks over the network using their computing resources. Later several other proof-based algorithms such as Proof-of-Stake (PoS), Proof-of-Elapsed Time, Proof-of-Space, and Proof-of-Luck were also proposed. Noticing blockchain's ability, many major organizations such as JP Morgan and IBM have begun to explore the technology. This gave rise to many other platforms in blockchain like private blockchain, consortium blockchain, etc. Unlike public blockchain, private blockchain allows nodes only from a single organization to involve in the verification process. In consortium blockchain, the nodes can be from different consortiums, but only permitted nodes are allowed in the network. Consensuses on these platforms are based on the results of majority. Hence, they are otherwise called as "voting-based consensus algorithm." Nodes following voting-based consensus algorithm can either be (i) a crush fault tolerance or

(ii) a Byzantine Fault Tolerance (BFT). Proof-based consensus algorithm can be applied even in private and consortium blockchains, and so is not limited to public blockchain.

4.2   PROOF-OF-X (POX) SCHEMES

4.2.1   REACHING CONSENSUS – THE ISSUE OF BYZANTINE GENERALS

Generally, Bitcoin consensus functions in a distributed way. There is a need for some kind of redundancy to guarantee fault tolerance, and the level of redundancy relies on the type of failures. For example, assume that three objects hold a value, and that one object fails and returns the same incorrect value. It is easy to find the failing object by comparing the results and decide on the correct value. So, the network requires $n \geq 2f+1$ in case of f failures. Nonetheless, failures can broadly be malicious or random. These failures are termed as Byzantine Generals Problem [6]. The real Byzantine problem is that n generals attempt to decide on a battle plan jointly by a messenger. Yet, f generals are traitors trying to break the deal. This problem is similar to an attempt to reach consensus by a distributed system. With consistent and trustworthy communication, Byzantine Generals Problem can achieve consensus [6]. The Fischer–Lynch–Paterson (FLP) proof demonstrates that no failures can be tolerated in asynchronous consensus protocol. Bitcoin uses randomness in mining process and readjusts it frequently. It does not allot any final or fixed peers for the validation process. Rather, it makes a thumb rule that transactions can be completed after six confirmations.

Another technology that directly takes the result of Byzantine agreement is Ripple. To build consensus, Ripple uses a group of reliable authorities. It uses a round-based consensus to make final decisions. The final decision is nothing but the current state of all accounts. $5f+1$ resilience is achieved in the Ripple consensus mechanism.

One more approach for consensus is the timing model. In synchronous message propagation, every message arrives after a time period. A message that takes a longer time is considered a failure. On the other hand, in asynchronous propagation, no decision is made on message delivery. Ref. [7] states that the timing model may not be possible in asynchronous models as messages usually take longer time to deliver.

In addition, the anonymous nature of Bitcoin makes Byzantine failure harder as it is possible for any attacker to introduce fake peers to divert the election process, thereby executing Sybil attacks. However, Nakamoto says that Bitcoin handles the Byzantine problem in a realistic way. Bitcoin assumes that an unknown-size synchronous network reduces the deterministic limit and embraces the final consistency. According to Ref. [8], under these notions, a fault tolerance of $2f+1$ is derived, where f is the sum of hash power of mischievous miners (Byzantine failure). It means that even if a network has malicious miners, it can reach consensus as long as honest miners hold more than half of the hash power. Bitcoin holds fault tolerance only if synchronous property is maintained. Therefore, information transmission between honest miners is important particularly when the hash power of malicious users reaches a threshold value, which may make the system delicate and insecure.

4.2.2   Proof of Work – The Question of Monopoly

The key component of Bitcoin is PoW. It is a random process of finding an answer to a riddle-like series of hashes. Predecessors of Bitcoin such as B-money, RPoW, Karma, and Bitgold already use proof of scheme in one or the other form. The objective of PoW in all these cases is to find a solution to a puzzle. Reusable proof of work (RPoW) follows a centralized approach in which PoW is reused as token money. As a token of acknowledgment, a server issues coins that are movable and reusable. Conversely, B-money uses a decentralized approach. The process is decentralized by choosing a transmission channel that is synchronous and unjammable. A transaction can be made by signing a contract and is informed to everyone in the network. A set of servers will keep track of the transaction in ledger. Similarly, Karma also follows a decentralized approach by having a bank-like setup. Yet, Bitgold follows the most advanced approach by linking PoW. The last entry in the chain then generates the next challenge and accordingly changes the difficulty. But Bitgold suffers Sybil attacks. Finally, using a refined PoW concept, Bitcoin combines resistance to Sybil attacks and coin-making process.

PoW works on the basis of "One-CPU-One-Vote." SHA-256 is used as a basic function by Bitcoin. Miners are revenue seekers. The mining cost is the cost involved in mining hardware and energy cost. Miners strive to find a solution for a mathematical problem as fast as possible. Miners use a normal CPU to solve PoW. But the speed of the CPU is limited. Therefore, each miner needs to find a faster solution to compete with other miners to generate more revenue.

Mining process can be extremely serialized. Hence, to do the repeated hashing operations in a faster and efficient way, Graphics Processors (GPUs) are used than CPUs. Bitcoin modifies the complexity (target value) in addition to the computing power. It is designed to maintain a target rate of ten minutes per block. The voting power per entity can be improved by using advanced mining equipment. This implementation undermines the PoW principle, and thus implies a risk. In fact, suppression of small miners undermines the democratic base. As a result, Bitcoin's confidence degrades. Actually, only the "rich get richer" in Bitcoin, i.e., wealth of wealthier users is growing faster than that of low-balance consumers.

During the early steps of PoW and before Bitcoin, the inequalities of systems were already identified as a potential problem. These are identified as memory-hard functions. However, it was proposed to use memory-bound functions to fix the situation. The solution is to integrate large quantities of (unpredictable) memory accessing operations into PoW calculations, making them the dominant variable. Therefore, the solution of the PoW is limited by disk access time, not CPU speed. Functions such as Scrypt and CryptoNight are explored in the context of Bitcoin. In a corresponding PoW scheme, memory-intensive operations are applied to the hashing operations. The idea is to develop a fairer distribution of power among the users in order to avoid a monopoly. The most popular alternative PoW scheme, Scrypt, enables specialized mining equipment to be used. In general, another fundamental drawback of PoW and Bitcoin, in particular, is that it consumes computational power with no intrinsic value. NooShare, Permacoin, and Primecoin change the situation by scheduling of arbitrary Monte-Carlo simulations as a PoW, finding long chains of prime numbers termed "Cunningham chains" and "distributing storage," respectively. Though these approaches provide some value addition apart from securing blockchain, it also has some difficulties like fine-tuning and non-reusability of earlier results.

4.2.3 PROOF OF STAKE – RESOLVING INCENTIVE PROBLEM

While Bitcoin clearly shows that a PoW-based currency is feasible, there are still weaknesses. Bitcoin's users pay the miners through an inflation process to protect the currency. Nonetheless, the future remains uncertain when the reward for the block decreases over time. Obviously, the block bonus halves about every four years (i.e., every 210,000 blocks). After halving the first reward, we can note a slight dip in the difficulty, suggesting that the miners leave the network. If the regulated coin supply continues as defined, the reward would be lesser than 1 BTC roughly in 2032, and it will come down to zero in 2140. This deflation, according to Ref. [9], is a process of self-destruction. It is putting the protection of digital currencies at risk by moving out the miners. Whether the transaction fees will be sufficient to account for the reward and provide the necessary motivation for miners remains unclear. The connection between the miners' strategy and the mining reward follows a game theory–based trade-off. Once a miner finds a solution, it must be propagated before the others declare the following block to receive the reward. The number of blocks does not affect the difficulty of the PoW, but a greater number increases the time required to reach consensus: The longer it takes for it to be distributed over the network, the more bytes the block contains. If there are more transactions in a block, it takes longer to verify their validity, but the higher the number of transactions, the higher the reward. Nash equilibrium states that a point where no individual can benefit by changing their strategy. Interestingly, in this Nash equilibrium, miners will not include any transaction in their blocks. This would certainly make Bitcoin completely useless, which also affects miners. The Nash balance shifts once the transaction fees increase or the reward for the block drops significantly. This implies that there may be an incentive for transaction fees.

There is, however, a problem with declining incentives that leads to the commons tragedy [10]. It is a gaming term which states that using individual, independent actions decreases the long-term gain of peer group by decreasing a common resource. Now let us look into the alternatives for PoW that allows us to revisit the basic requirements that must be met: First, block generation must in some way be "expensive," and individual miners must not be able to acquire an excessive potential to mint coins. Second, consensus must ultimately be achieved; a common rule must be established for resolving forks and determining of main blockchain. And finally, it should be non-fraudulent.

Coin age is considered as a possible alternative to PoW. Coin age is the amount of times the currency has been inactive. Assume if user A sends two coins to user B, and user B retained the coins for 80 days, then the coin age is 160 coin-days. When user B expends the two coins, the coin age accumulated is damaged. This method of using coin age to describe the incentives is called Proof-of-Stake (PoS). It is applied in Peercoin. A PoS block mining needs a coinstake block (similar to Bitcoin's Coinbase transaction). Owners send their coins to themselves in a coinstake transaction and add an already-defined percentage as their reward. To successfully mine a block, a hash value lesser than or equal to a target value is needed. The difficulty is determined individually unlike PoW. Since the hash is based on static data apart from a timeframe, miners cannot use their computing power to solve the puzzle faster than others. There is no nonce that can be changed. Instead, the time-stamp changes every second, and miners are given a new chance to discover the response. When a solution is found, the coinstake transaction block is announced to the server. The miner collects the reward provided by the coinstake transaction, thus resetting the coin age at the same time. Subsequently, a new coin age can, of course, be gathered again, slowly increasing the chances of solving the puzzle next time.

Coin age can be compared with PoS, in terms of computational power in PoW. But the key difference is that the "power" is independent of the computer power. Thus, PoS offers a response to the accusation that PoW is wasteful of resources. Unlike Primecoin, which tries to put an inherent value in PoW, the high energy usage is greatly reduced in PoS. Furthermore, miners kill the age of the coin by demand-ing the reward; they don't hold it for the next round. Such properties decrease the threat of monopoly that exists in the tragedy of the commons problem. Equal voting power is distributed. Therefore, "rich gets richer" is balanced by "poor gets richer," which means that each participant can provide PoS, thereby helping to protect the blockchain and also earning a percentage of incentive for keeping them. PoS cur-rency is seemed to be more expensive to use. PoS declares a chain with the highest total of destroyed coin age as the main chain in contrast with PoW where the longest blockchain survives. PoS is added for practical purposes to supplement PoW and to become the major factor if incentives subside in PoW. Nonetheless, altcoins such as Nextcoin also have a complete PoS platform. Derivatives such as transactions as proof of stake (TaPoS) and delegated proof of stake (DPoS) are intended to mitigate the problem of monopoly while making the system safer by obtaining votes from a wider range.

4.2.4 PROOF OF ACTIVITY – PROMOTE ACTIVE INVOLVEMENT

PoS offers an alternative to PoW, but it has a number of limitations. According to Ref. [11], when the reward for a block falls and the transaction fees are to be assessed it leads to a situation called tragedy of commons where a miner has o maintain a reward for every payment, including some charges for himself. The miner will not communicate the payment ultimately to reduce rivalry. Peercoin eliminates the fees to remove the incentive for non-cooperation, rather than paying the miners transac-tion fees. However, there are certain disadvantages to PoS. This depends completely on the age of the coin, which can only be obtained by keeping the coins and can be claimed in a coinstake. Expending coins on daily transactions, i.e., regularly assign-ing coins to others, also kills the coin's age as these coins are not included in the PoS lottery.

The biggest weakness, though, is that coin age accumulates even if the node isn't connected to the network. In other words, it is sufficient for a node to be online for some time and be offline later. This behavior will result in a more bursty distribution of rewards than with nodes remaining online all the time, but most likely investors will embrace that. The lack of a sufficient number of online nodes also makes threats worse. Any incentive structure aims to encourage in some way or the other must pay attention to Sybil attacks.

The advanced compensation scheme of PoS indicated that increased activity would lead to a healthier economy. He [11] addresses the problem that coin age is a linear time variable. In fact, to ease some of the mentioned problems, Peercoin uses upper and lower limits of coin age. For example, switching the increment function to an exponential decline function would have a profound effect. In such a system, the rate of increment of coin age is decreasing over time and converging linearly to zero. Parameterization of the decline constant allows explicit definition of the feature's half-life time. It improves the incentive: A fresh coin accumulates coin age much faster up to a fixed value. This simple idea is called PoS velocity and is implemented at Reddcoin.

The solution to this is to specifically reward positive colleagues for their effort. The idea is to giveaway a part of the PoW block incentive among all the active nodes, while their stake will measure the amount of prizes, i.e., their chances of winning. So, it's a mix of proof of function and PoS. Miners mine empty blocks. If the problem of PoW is solved, it will be broadcasted in the network as before. Anyone who receives the block derives from it N deterministic numbers of pseudorandom tickets. Using their respective private key, the first $N-1$ users sign the block and sends the signature. If the N-th most fortunate stakeholder sees the block, it produces a tag, containing the block, all transactions, $N-1$ signatures, adds its own signature and transmits the wrapped block. If both the block and the fortunate investors are correct, others will find it as a genuine extension of the blockchain. The transaction fees finally will be shared between the investors and the producers.

The basic concept is to reward active online peers. If any stakeholder is offline, he can't reply and include his signature. Thus, the block cannot be completed. Another miner will solve the work proof at some stage, drawing specific N stakeholders. The concept of changing the hash rate and the percentage of participating peers is known as Proof of Activity (PoA). This approach rewards participating stakeholders, instead of punishing passive stakeholders. PoA enhances security: An invader requires a certain amount of stake in addition to a huge amount of computational power to execute a double spending. PoA in some aspects is an essential component of peer-to-peer networks. For example, the BitTorrent protocol features a built-in direct reciprocity reward mechanism ("tit-for-tat"). This approach's efficacy has been demonstrated many times. It is to be noted that the Bitcoin network follows a different objective, i.e., quick block propagation rather than file sharing. Hence, PoA can be conceived as indirect reciprocity.

4.2.5 PROOF OF ELAPSED TIME

Hyperledger Sawtooth [12], an open-source project, uses its own consensus algorithm called Proof of Elapsed Time (PoE). It is developed in Intel's Software Guard Extensions (SGX), a Trusted Execution Environment (TEE). A trustworthy voting system based on the SGX assists in selecting a validator to publish a new block. PoE is a lottery-based consensus algorithm, but it fails to fulfill the need of solving complicated, expensive mathematical puzzles. Nodes request for a wait time from a trusted method within the SGX in the Sawtooth blockchain network. The winner is chosen based on the shortest wait time. Another trustworthy method testifies that the validator actually waited a certain amount of time before a new block was released. Therefore, the proof of the validator is selected after the allotted time has elapsed.

4.2.6 TENDERMINT

Tendermint is a kind of Practical Byzantine Fault Tolerant (PBFT) consensus algorithm. Similar to PBFT, it provides a n 3f+1 fault tolerance. It uses stake proof in conjunction with PBFT concepts to ensure safety, high throughput, and a processing time of 1–3 seconds at low blocks. Whereas a leader node is chosen pseudorandomly in PBFT, Tendermint embraces the lottery-based PoS properties and chooses the lead node with a probability proportional to the network share of stakeholders. Tendermint conducts several rounds of voting after member selection to reach consensus on addition of a new block to maintain 100% uptime. It needs a majority (about two-thirds) of its validators, and if more than one-third of the network goes offline, the progressing of network may stop and lose productivity. Transactions are organized, and if less than one-third of all verifiers are unreliable, it provides a safety guarantee that there are no competing blocks created and no forks in the blockchain present. Tendermint is compatible with public or private chains, but it does not enjoy the same level of scalability as PoW or PoS.

4.2.7 FEDERATED BFT

Ripple and Stellar implementation of blockchain expanded the conventional BFT for situations involving a node federation or consortium. Ripple consensus starts with a distinct node list (UNL) containing a list of active validator nodes in the network. Each node has a node list of 100+ nodes, and each list will overlap by atleast 40% with the lists stored by other nodes. Ripple performs several rounds of voting in which the nodes compile the transactions into candidate blocks and broadcast them to the other nodes in their list. Later, the nodes send the votes on every candidate block. Each round of voting lets nodes optimize their candidate block, and a new block is introduced to the list once an 80% majority vote is received. Thus, by conducting multiple rounds of votes, Ripple provides a sensitivity of $n=5f+1$ fault.

In blockchain networks, Stellar presents the concept of quorums. A quorum is a set of nodes used to reach consensus. A node in such a network will form part of multiple quorum slices in which each slice of quorum safely reaches consensus by voting. As quorums and quorum slices are permitted to converge within the main Stellar network, Stellar allows open node participation in different sub-networks. Since Stellar chooses for security over liveness property, the blockchain will not advance until the malicious activity is resolved in the event of a malicious behavior in the network. Stellar offers versatile trust and low latency as it is computationally affordable. Quorums include a limited set of nodes that exchange voting messages.

Table 4.1 presents a comparison of the consensus models that are discussed so far.

TABLE 4.1

Comparison of Blockchain Consensus Mechanisms

	PoW	PoS	PoET	BFT	Federated BFT
Blockchain category	Permissionless	Permission and permissionless	Permission and permissionless	Permission	Permissionless
Transaction decision	Likely	Likely	Likely	Instantaneous	Instantaneous
Transaction speed	Less	More	Medium	More	More
Requirement of token	Required	Required	Not required	Not required	Not required
Participation cost	Required	Required	Not required	Not required	Not required
Network scalability	More	More	More	Less	More
Trustworthy model	Untrsuted	Untrsuted	Untrsuted	Semi-trusted	Semi-trusted
Tolerated power of adversary	≤25%	Depends on procedure used	Unfamiliar	≤33%	≤33%

4.3 PERFORMANCE AND SCALABILITY IN CONSENSUS ALGORITHMS

To consider the transmission speeds of nodes within the network, permissionless blockchains are expected to have slower block formation speeds. On the other hand, permissioned blockchains are much less latency but suffers a serious problem of scalability. The network overhead generated by voting systems limits permissioned blockchain to scale in a limited range, i.e., only hundreds of nodes. The worst-case complexity for permissioned blockchain is O^1N2^o, and O^1N^o is the worst-case complexity of permissionless blockchain. Therefore, from PoW consensus to PBFT consensus, there is a drastic trade-off between efficiency and scalability. Permissionless blockchain is better for Internet of Things (IoT)-related applications due to its principles of anonymity and decentralization, whereas because of their higher degree of control and authorization, and permission granting capabilities, permissioned blockchain is more appropriate for enterprise solutions.

CONCLUSION

The various consensus mechanisms are briefly discussed, and these protocols are analyzed and compared on various aspects. The performance and scalability aspects of these algorithms are also studied and presented.

REFERENCES

1. M. J. Fischer, N. A. Lynch, and M. S. Paterson, "Impossibility of distributed consensus with one faulty process," *Journal of the ACM (JACM)*, vol. 32, no. 2, pp. 374–382, 1985.
2. D. Dolev, C. Dwork, and L. Stockmeyer, "On the minimal synchronism needed for distributed consensus," *Journal of the ACM (JACM)*, vol. 34, no. 1, pp. 77–97, 1987.
3. https://www.ethereum.org.
4. S. Popov, "A probabilistic analysis of the Nxt forging algorithm," *Ledger*, vol. 1, pp. 69–83, 2016.
5. A. Back, "Hashcash: a denial of service counter-measure," 2002 [Online]. Available: ftp://sunsite.icm.edu.pl/site/replay.old/programs/hashcash/hashcash.pdf.
6. L. Lamport, R. Shostak, and M. Pease, "The Byzantine generals problem," *ACM Transactions on Programming Languages and Systems*, vol. 4, no. 3, pp. 382–401, 1982.
7. M. K. Aguilera, "Stumbling over consensus research: Misunderstandings and issues," in Bernadette Charron-Bost, Fernando Pedone and André Schiper (eds.), *Replication*. New York, NY: Springer, 2010, pp. 59–72.
8. A. Miller and J. J. LaViola, Jr., "Anonymous Byzantine consensus from moderately-hard puzzles: A model for bitcoin," Computer Science, Univ. Florida, Gainesville, FL, USA, Techical Report, April 2014 [Online]. Available: http://tr.eecs.ucf.edu/id/eprint/78.
9. N. T. Courtois, "On the longest chain rule and programmed selfdestruction of crypto currencies," Computing Research Repository, Technical Report abs/1405.0534, 2014.
10. G. Hardin, "The tragedy of the commons," *Science*, vol. 162, no. 3859, pp. 1243–1248, 1968.
11. G. Pickard et al., "Time critical social mobilization: The DARPA network challenge winning strategy," Computing Research Repository, Technical Report abs/1008.3172, 2010.
12. N. Szabo. "Bit gold." December 12, 2018 [Online]. Available: http://unenumerated. blogspot.de/2005/12/bit-gold.html.

5 Blockchain
Introduction to the Technology behind Shared Information

Naseem Ahamed
ICFAI Business School

CONTENTS

5.1 INTRODUCTION TO THE BLOCKCHAIN TECHNOLOGY

Blockchain is the technology that enables its participants to verify existing blocks of information and add new ones based on a consensus algorithm. It generates time-stamped chunks of information encapsulated in blocks secured by hash signature and is immutable by nature. Blocks after blocks get added in the network as it increases in size with the addition of new participants and increased frequency of usage. A user has to download the blockchain architecture on his computing device either partially or in its entirety depending on his intention to use the blockchain. Once an information is verified and added to a block, the information is open to see for everyone in the blockchain, and hence it is popularly also known as the distributed ledger system. The blockchain technology can be used for multiple purposes, but it is widely known as the technology used for transaction of cryptocurrencies (Bitcoin being one of the most popular cryptocurency). In popular culture, the distinction between blockchain and Bitcoin [1] is vague as they are often used interchangeably. However, to say that blockchain and Bitcoin are the same is akin to saying that electricity and light bulb are the same. It is common knowledge that electricity is the technology that has umpteen applications, a light bulb being just one of them. Similarly, blockchain is the technology with various applications including Bitcoin.[1] With the passage of time and acceptability among user groups, other currencies entered the fray with more applications such as smart contracts and other distributed applications.

5.1.1 BLOCKCHAIN: TYPES OF BLOCKCHAIN

Another misconception regarding blockchain at large is that there is one big comprehensive global blockchain consisting of information of all transactions. However, this is far from truth as there are multiple blockchain networks. The content on those networks would be determined by the users maintaining them, e.g., the blockchain network trading Bitcoins is a network for peer-to-peer transaction of Bitcoins only. Ethereum [2], however, is a different blockchain platform [3] that allows users not just to transact ether[2] but also to make and maintain smart contracts[3] among the

---

[1] Bitcoin is the name of an extremely popular cryptocurrency. A cryptocurrency is a virtual asset that can be used in place of fiat currency as a medium of exchange. It uses advanced cryptography to escape alteration and duplicity, and ensure security.

[2] Ether is the name of the cryptocurrency used on the Ethereum platform. It is an alternative currency to Bitcoin.

[3] A smart contract is a misnomer because it is not a contract in the real sense of the word. It is a condition-based document that executes predetermined actions based on fulfillment of the mentioned conditions.

A very simple example in the field of stock trading would be that an investor sets a threshold floor price for a share. No sooner than the stock price falls below that threshold price, the sell command would automatically get executed and the proceeds of the sale would get deposited in his trading account.

The next example from the field of supply chain would be that the supplier's account would get automatically credited once the shipment arrives inside the warehouse and invoice is signed.

users on the network. There can be multiple blockchains existing simultaneously with information on them. There are primarily two types of blockchains, namely public blockchain and private blockchain.

5.1.2 PUBLIC BLOCKCHAIN

A public blockchain, as the name suggests, is a network that does not require any permission for new participants. It has no restriction whatsoever, and anyone can join and contribute (through validating transactions with proof-of-work method or otherwise) to the growth of the blockchain. A new participant just needs to download the existing blockchain and join the network. The incentive for authentication on public networks is nominal fees for mining, i.e., miners who authenticate Bitcoin transactions receive some Bitcoins as a fee.

5.1.3 PRIVATE BLOCKCHAIN

A private blockchain, however, is a network that does require invitation from the administrators of the network. A newly joined participant can be given either full or partial access to the network. A full access would mean that the new participant can not only view the information but can also authenticate subsequent information and add new blocks in the process. A partial access might mean that he does not have authentication provision. Hence, in a private blockchain, the degree of access to new participants is decided by owners of that blockchain. Such blockchains are more commonly used within organizations among employees.

5.2 ISSUES WITH THE CONVENTIONAL SYSTEM: TRUST AT STAKE

In its current form, the world of finance is based on an architecture where intermediaries execute transactions between interested parties. These intermediaries such as banks and financial institutions are bound by fiduciary responsibilities for their customers. It is because of the nature of this fiduciary responsibility, and it is imperative for the participants to have faith in the intermediary. The intermediary executes the transaction for a small commission/fee. However, many a time, they deviate from their fiduciary responsibility and act in self-interest either through rent-seeking behaviors of top executives or otherwise. The world witnessed the fall of once mighty Lehman Brothers. Other giants felt the heat including American International Group (AIG), Merill Lynch, Freddie Mac, Fannie Mae, and Royal Bank of Scotland, to name a few [4]. There was a growing sense of mistrust among investors over the non-fiduciary conduct of financial corporate bodies as depositories or intermediaries for their financial transactions. A similar situation resulted before in the enactment of Sarbanes-Oxley Act of 2002. Although blockchain as a technology was being experimented with and developed for decades, it was only after the white paper by Satoshi Nakamoto [5] that the technology of blockchain entered public conversation. The timing of the paper coincidentally matched with one of the worst financial crises that the world had seen in the past decades.

5.3 FIXING THE ISSUES WITH THE CONVENTIONAL SYSTEM: THE WHITE PAPER BY NAKAMOTO

The article by Satoshi Nakamoto had the exact answer to the question that people had in their minds. Often, financial intermediaries have been perceived in the minds of investors as a "necessary evil," i.e., they don't trust them fully, but there are no other alternatives to their services. Not only did Nakamoto propose an alternative to these financial intermediaries through his peer-to-peer network of electronic cash system, but he also asserted that it would not incur any additional service fee or commission for investors for its services (Figure 5.1).

5.3.1 EMPOWERING PARTICIPANTS BY DECENTRALIZING THE ARCHITECTURE

It was a win-win situation for the investing community where they can have a safe and secure corridor for executing transactions and save additional service-related expenses. The idea quickly spread, and proponents of blockchain asserted that blockchain technology has the answer to the existing problems of our current financial system. The current financial system is a centralized system that is the perfect condition for corruption to breed. Also, having all the data stored in a single location makes it vulnerable to theft, manipulation, etc., whereas blockchain is a decentralized system by design. It is like a time-stamped public ledger that runs on multiple systems and is immutable. Any individual or group trying to hack the system would

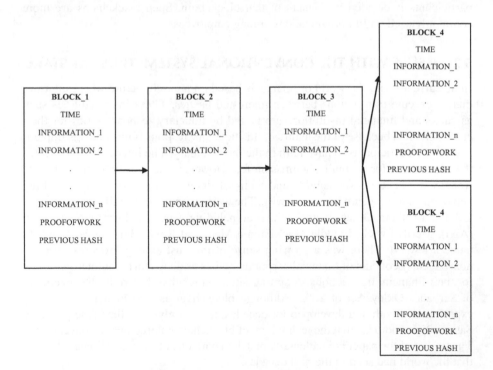

FIGURE 5.1   Simplified structure of the blockchain architecture.

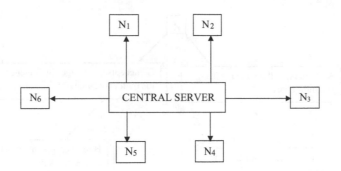

FIGURE 5.2   Structure of the traditional centralized system.

require to have consensus with remaining nodes, which is very less likely on its own. In addition to that, the amount of money, computing power, and other resources (memory and otherwise) required for such a concerted effort would not be worthwhile for the hackers (Figure 5.2).

So, by design, the decentralization of control makes blockchain less vulnerable to attacks, and failure of one node does not mean the collapse of the entire structure.

5.3.2   AUTHENTICATION OF BLOCKS THROUGH CONSENSUS ALGORITHM

The blockchain technology is very democratic with respect to the existing system. In a blockchain, blocks of authenticated and approved transactions (encoded in crypto form) are added to the existing blocks with hash authentications (Figure 5.3).

Bitcoin uses the proof-of-work[4] system to authenticate a transaction before adding it to the existing blocks. A proof of work is a type of consensus algorithm to determine the node that has spent computation power to solve a mathematical riddle. The winner node will get to add the new block and would be rewarded. Each block is joined with its previous block with the help of cryptographically authenticated hash of the new

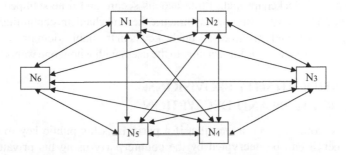

FIGURE 5.3   Structure of the decentralized system.

---

[4] Proof of work system is a system to prove sufficient computation executed by the miner before authenticating a transaction. Other methods for proving computation such as proof of stake, proof of elapsed time, proof of capacity, and proof of deposit are also there but Bitcoin uses proof of work for its consensus algorithm.

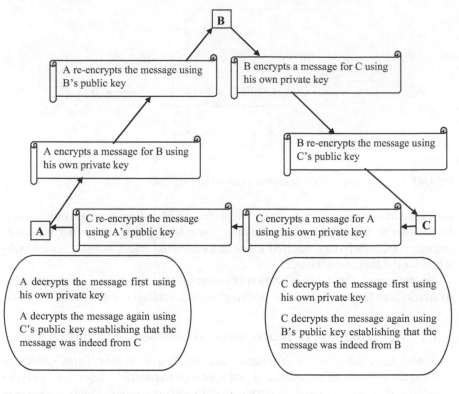

FIGURE 5.4   Communication between nodes.

block. New blocks are continuously added to the existing ones after consensus by the network, and thus it forms a chain structure deriving its name blockchain (Figure 5.4).

5.3.3 ENHANCED SECURITY CHARACTERISTICS

The existence of blockchain makes transactions secure and time-stamped [5]. Users can encrypt and decrypt using an asymmetric key mechanism containing a pair of keys, viz. public key and private key. A public key is in the public domain and known to all, whereas the private key is known only to the users with whom one wants to interact.

5.4   CONFIDENTIALITY PROVISIONS: ENCRYPTION AND DECRYPTION

The process works in such a way that if a party uses his public key to encrypt a message, then it can be decrypted by the counterparty using his private key and vice versa. So, to ensure confidentiality, the first party encrypts his digitally signed message to the counterparty using his private key first and then re-encrypts the message with the counterparty's public key [6]. This process of encrypting the message twice using private and public key serves dual purpose. First, it provides two layers of encryption making the message more secure. Second, every user needs to have two keys only, i.e., public key and private key. In order to decrypt the message, the counterparty has to use his private key and then again decrypt it using the sender's public key verifying the sender's authenticity. Each user would have to have thousands of keys, one for each user if not for these public and private keys.

5.5   SELF-SUSTAINABLE ARCHITECTURE: INCENTIVIZATION TO MINERS

Specifically in the case of Bitcoin, miners are rewarded with Bitcoins for executing this important task of authentication by proof-of-work system. This serves as an incentive for them to spend time, effort, and resources (memory, electricity, CPU with high processing power, etc.). The total number of Bitcoins has already been fixed at 21 million. In that case, as more and more transactions take place and the usage of Bitcoins becomes more acceptable, the mining fee for the miners (paid in Bitcoins) [7] would increase in value. This is a simple case of "Demand and Supply" in economics where the rarity of a good determines its value. As Bitcoin becomes more and rarer, it would command more value and vice versa. It also raises a pertinent question as to what would happen if the entire pool of 21 million Bitcoin [6] is mined.[5]

5.6   SECURITY FEATURES: DIGITAL SIGNATURE

The blocks in the blockchain are secured using hash signature that is a digital identification unique to that concerned block only. Each block has the hash signature of the previous block, and only after authentication by the consensus of the nodes, new blocks of transactions are added to the existing chain. It is because of these hash signatures that the blockchain is so difficult to alter. If someone tries to alter or attack a block, then its hash signature would mismatch and the following block would not accept the previous block. So, the attacker would have to alter the next block, which would result in rejection from the subsequent block and so on. It would require a concerted effort from a malicious majority of the nodes to change the data in one block and get a consensus and then keep doing so for all the blocks in that chain.

5.6.1   HASHING

A hash function is a complex mathematical function that processes any arbitrary input of any length into an output of a fixed length. Different software can be used for the

---

[5] It would take some time for such a situation to come because Bitcoin is not fully accepted as a means of exchange in our society. There is resistance from various quarters such as government, financial houses, and many investors. Governments directly and indirectly control the amount of money in an economy through its monetary and fiscal policies and intervening in situations when some sector needs special attention. Financial institutions are averse to new-age cryptocurrency because it is a direct assault on their business model. Finally, many investors are cynic because the technology is relatively new and they are hesitant to commit to a completely unknown system of transaction. Cyber security is another concern with them, and finally in case of a failure in traditional financial system, the culprits can be taken to task by the court of law, but in blockchain, there is no one to point fingers at.

With that backdrop, even if all the Bitcoins get mined, there are two possible scenarios. First, the peers on the network can unanimously decide to increase the overall supply. Second, miners would start getting their fee from the users trying to execute transactions.

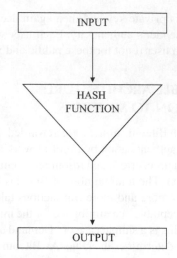

FIGURE 5.5 Process of hashing.

process of hashing; notable among them are Message Digest (MD family of software) and Secure Hash Algorithm (SHA family of software). SHA-256 takes an input of arbitrary length and processes it through the hash function to generate an output of 256 bits. Any change whatsoever in the input, no matter how small, would result in a totally different output of 256 bits. SHA-512 does the same job, but the length of the output is 512 bits. Hashing is a one-way traffic where every time the same input would result in the same output, but one can't derive the input from its corresponding output (Figure 5.5).

5.6.2 Consolidating Multiple Hash Signatures

A block in a blockchain might contain information of multiple transactions with each of them having their unique hash signature identity. These multiple hash signature identities within a block are aggregated to form a totally new hash signature identity till the root hash signature unique identity is reached. This unique signature identity serves as the hash identity for the whole block, and it is this hash signature identity that is used for linking of this block to the following block (Figure 5.6).

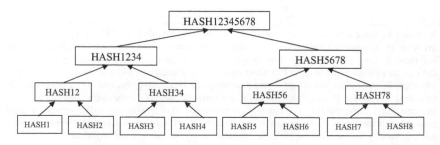
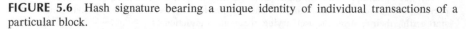

FIGURE 5.6 Hash signature bearing a unique identity of individual transactions of a particular block.

INPUT	HASH FUNCTION	OUTPUT
I am Tabasseem	SHA-256	56CLO97J976FR67YU8I076YUHHTGGG56YJJU53 1J8UUG67809YTFDE567JLU8I9R53XZSDFVMKJ0 O9Y7UT6Y64DE4FGHB654CM8K74KJU58GFR53
IamTabasseem	SHA-256	H76FIOFG54WCFG708D85DKDIIE90EJDKD85JUI OD455FRF485765KJ6D42DHNMLDXS148LJ78GH 485KLJH78H4SD43DF66JU98JUJ09894LKOIU4759
I am tabasseem	SHA-256	1WFR56YHGDCG8907YHU8754RFTYH765BHUJ8 765BMK4D4F8R5FG5G1D7S1EF4F7F85DFF5F7F4 45UIIO098U7F65HY776TGHY762AS3D5CG654YU
I Am Tabasseem.	SHA-256	I90O76TR54DFGT5423FGH768JJKMNG7K58L5J1J 1H86Y7D63X5D76FGHY7864GBHJKIO09G568GR LP0OO48KH78F54D5F78H9H0D56V7NB78X56MK
iTabasseem	SHA-256	H8797HY73ZA24XM0KO980HU78865FRTGBHD4 566567TYRFH653ZS234FV789IOURBNMJH56FDI 4KOL586HYF56DC67VH90ZA32CV5G6LK9076D4

FIGURE 5.7 The process of hashing through an algorithmic software.

A simple hypothetical example of hashing is given in the table shown in Figure 5.7, where five different versions of the same message with slight modifications are entered into a hash function (SHA-256) for processing. Although the output generated in each case is vividly different from each other, the length remains exactly the same.

5.7 ROAD AHEAD

Blockchain, similar to any other technology in its early stage, has been greeted with both enthusiasm and apathy. On the one hand, the supporters of the blockchain technology believe it to be the latest wave of revolution in the information and communications technology with the potential to radically impact multiple sectors. On the other hand, there is a group of naysayers who are of the opinion that blockchain technology is the latest buzzword which would never achieve its purported objective at best, and would be a convenient tool for illegal activities at worst.

5.7.1 IMPACT ON EXISTING INDUSTRY: A NEW WAVE OF TECHNOLOGICAL REVOLUTION

Advocates of the blockchain technology are of the opinion that this is the biggest revolution in the field of computer and technology since the advent of the World Wide Web (www). They claim that it has the potential to change every major industry as we know it. It is not a marginal increment in the existing technology, but it is here to devour the existing technology and create things in a totally different way.

Many organizations that serve as middlemen or intermediaries between two parties in a transaction would be rendered irrelevant. Large organizations spend millions in improving service quality, grievance redress mechanism, advertising and promotions, etc. and work toward customer satisfaction to develop trust among people. Customers value this trust, and it is an important determinant when they are choosing their intermediary such as stockbroker company, online shopping company, policy aggregator, and bank. The brokerage company helps investors match with the stock/debenture of their choice; online shopping aggregator helps customers match with the seller selling their favorite product; policy aggregator helps customers match with the company that sells policy which suits the investor needs; and banks both serve as market maker and custodian of customer's deposits, and also transfer money from one party to another within and outside the country. All these intermediaries charge either a fixed sum or a portion of the transaction value as service fee for executing the transaction. The wide acceptance of blockchain technology and transition of business toward the same would result in complete rout of market makers and fee-/commission-based business models as all the information would be available in the public domain. Some of these intermediaries are already aware of the possibility of a tectonic shift in their business and hence are trying to adopt the technology at the earliest to fit their business model accordingly [8].

5.7.2 IMPACT ON EXISTING INDUSTRY: DOOMED TO FAIL

Cynics of the blockchain technology [9], on the other hand, opine that the hype created around the blockchain technology is synthetic in nature, and blockchain itself is a buzzword at best and a convenient tool of transaction in the hands of smugglers at worst. The naysayers assert that apart from being a relatively new technology that has not been tested successfully in a comprehensive manner, blockchain is a complex architecture to understand for beginners. Many governments, apprehensive about its impact on the economy coupled with loss of monetary control on the same, have either decided to ban its mining, usage, and possession altogether or stopped its use for some time to study it first. The wild volatility of cryptocurrencies could also pose grave uncertainty for that economy, which investors might want to avoid. The capacity of blockchain to handle user traffic at its current speed is very limited. Currently, each block takes about eight minutes to get authenticated and added in the chain, which needs to be reduced significantly with the improvement of technology. Security/privacy of the information on the public blockchain is another concern where through it is definitely difficult to attack the structure and make modifications, it is not impossible to do so. The intricacy of how malicious nodes would get a majority control of the network is not known clearly; however, if it does happen, it would be the end of it because majority consensus can be achieved by malicious nodes. Another issue with blockchain technology in terms of cryptocurrency is that the sender and receiver are anonymous. The information about transaction, its magnitude, time, etc. would be there on the block, but identification of the parties is anonymous. Hence, it can be used for illegal purposes very easily without proper regulation and monitoring agency. Finally, the system of using computational power for proof of work to validate a block requires a lot of energy, and it is totally unproductive given the objective to execute transaction at the earliest.

5.8   APPLICATIONS OF BLOCKCHAIN TECHNOLOGY

Applications of the blockchain technology are as diverse as it can get. Blockchain can revolutionize almost every sector of an economy as it is a very generic technology that can be adapted to suit the needs of a given sector/company. Few of the applications of this technology in different sectors are mentioned below. This is definitely not an exhaustive list, and many more uses, both major and minor, of this technology can also be explored.

5.8.1   PRODUCT AUTHENTICATION

This is one area in retailing where end customer can get full life history of the product that he intends to purchase. This history includes details right from the manufacturing process to where the product got quality clearance, to packaging and shipping so that the customers can be sure about the authenticity of the product. In this day and age when fake and counterfeit products resemble the original ones, it can be very hard to differentiate between the two. Customers can make informed decision based on their choice and need. It will also help customers know the age of the product[6] as all the information is available on the network.

5.8.2   CREATIVE INDUSTRY AND ACADEMICS

In the fields of creativity (such as art, music, and allied areas), intellectual expression (such as books), and intellectual property, the blockchain technology can serve as a big boon. The conventional system works in such a way that a major portion of the proceeds from the sale of music, books, etc. goes to other parties than the creator. The creator gets their royalty proceeds which is a small portion of the actual proceeds. Putting your product directly on the blockchain network would enable to create a direct link between creators and customers. The product can then be linked with smart contracts where it would automatically credit the proceeds from the sale of the product from the blockchain. The smart contracts can be designed in a way that would not only make the product secure but also more readily accessible to anyone who wants it with instant payment to the creator. Imogen Heap put her song "Tiny Human" on blockchain and has been a vocal advocate for this technology.

5.8.3   VOTING AND GOVERNANCE

One of the best uses of the blockchain technology given that it runs on the principle of consensus is to strengthen democracy *per se* by putting the electoral process on blockchain. The world over, there are allegations of influential political entities of tampering with the electoral process and rigging the election in favor of their candidate. In a democracy, the process of election is crucial because that is when

---

[6] Some products are valued for their age such as basmati rice and wine. Others are valued for their geographical origin such as Italian leather, Carrara marbles, Aligarh locks, and Muzaffarpur litchi. Having all the relevant information on the blockchain network helps the customer make a preferred decision.

people elect their representatives to voice their issues on larger platforms. A compromised election means that the people's representation also gets compromised. If the electoral process is put on blockchain, then the details of the event would be running on multiple nodes simultaneously. Everything ranging from the contestant's profile, to address of booth, to the number of voters in that constituency, to number of votes garnered by each candidate would be reflected on multiple nodes simultaneously. This digression from the centralized system would lead to more democracy in the democratic process. More centralized control over elections would lead to more vulnerability to the whole electoral exercise. However, it wouldn't be wise to shift to blockchain overnight in matters as important as elections because the pitfalls of such a transition are still unknown. Few pilot projects should be run first to assess the efficiency of the technology, readiness of voters, digital infrastructural issues, etc. before expanding it in a phase-wise manner.

5.8.4 Storage and Movement of Assets

All asset classes ownerships, i.e., real and financial assets such as land entitlements, stocks/debentures of companies, and patents, can be put on the blockchain. One issue that cryptocurrencies faced earlier was the problem of double spending. Double spending would lead the currency to become unusable because it makes no sense if one pays someone Rs 10 and he still has it with him. Once a party spends a certain amount, he should lose the hold of that and the counterparty should receive the same. Bitcoin overcame this double spending problem through time-stamping each transaction and passing the information of all the transactions on every node. Similar to Bitcoin, any asset of value can be put on a blockchain. It would help rightful owners escape from the scourge of document counterfeiters. Not only would it be helpful to rightful owners to obtain loans against their asset but also save scarce resources unproductively wasted in years of litigation. The advantage of this information being on blockchain is that not only it holds the name of its current owner/ custodian but one can go back and trace the previous owners as well. All change of hands of any asset can be well preserved in the blockchain.

5.8.5 Link Service Provider Directly with Customers

The cab aggregator service, policy aggregator service, hotel aggregator services, etc. charge a portion of the total amount paid by the customer as service fee, which provides for the maintenance of their infrastructure digital and otherwise, labor charges, etc. However, if all the information is available on the blockchain network, then the need of these aggregator intermediaries would go away. Blockchain can facilitate a direct link between service providers and customers benefitting both of them.

5.8.6 Blood Group and/or Organ Donor List

A publically available ledger of people with their blood groups on the blockchain network itself would be a lifesaver. Patients need blood all the time for various reasons mostly those with severe blood loss in an accident. Some blood groups are hard to find because of its rarity; hence, such a list would help locate donors nearby. A similar list of organ donors and receivers can be put on the blockchain to help execute a match.

5.8.7 ARRESTING DISTRIBUTION LEAKAGES

Many governments run welfare schemes for noble causes such as poverty alleviation, food security, and help in times of natural disasters, but a lot of their benefits never reach the intended beneficiaries. The blockchain technology can be a great way to improve public distribution system and arrest leakages in it. A lot of such schemes don't achieve the intended result just because of a broken distribution system, and at the same time, it leaves affected people with a sense of dissatisfaction and anger with their political dispensation.

Blockchain as a technology is here to stay and its application, while might not be used on a mass scale today but with improvement in the product and more awareness, would eventually appeal to users. It would certainly fail in some pilot projects and might achieve modest success in others, but with the passage of time and improvement in technology, and increase in computation power, it will ultimately make lives easier.

CONCLUSION

The rise of Bitcoin has been phenomenal within a short span of time where the value of one Bitcoin reached close to a million and a half U.S. dollar at its peak. After witnessing the success of a crypto-based digital currency, other alternative crypto coins entered the fray too. All this currency does is removing the need to trust other parties/intermediaries involved in the transaction. It is a digital architecture with peer-to-peer cash transfer system embedded in it. Everyone on the network is a witness to the transaction; hence, it is available in the public domain with almost no likelihood of alteration. The technology behind this system of financial transaction *sans* intermediaries is called blockchain. The blockchain technology with its advantages promises to touch almost every sector ranging from retail to medical. It is gaining a firm footing in many sectors, while other sectors have adopted a wait-and-watch policy for some time. Only time will tell how big a disruption would be caused by the blockchain technology, it is interesting to note that it has already come a long way in a span of a decade.

REFERENCES

1. A. Antanapoulos. *Mastering Bitcoin: Programming the Open Blockchain,* O'Reilly Publications, California, 2017.
2. G. Hurlburt, and I. Bojanova. "Bitcoin Benefit or Curse?," *IT Professional,* vol. 16, no. 3, pp. 10–15, 2014.
3. D. Vujicic, D. Jagodic, and S. Randic. "Blockchain Technology, Bitcoin, and Ethereum: A Brief Overview," *2018 17th International Symposium* INFOTEH-JAHORINA (INFOTEH), East Sarajevo, 2018, pp. 1–6.
4. M. Leising. "Wall Street Embraces Blockchain as the Future," *Sydney Morning Herald,* August 2015.

5. S. Nakamoto, "Bitcoin: A Peer-to-Peer Electronic Cash System," White paper, 1–9, 2008.
6. A. Laska, B. Johnson, and J. Grossklags. "When Bitcoin Mining Pools Run Dry: A Game-Theoretic Analysis of the Long-Term Impact of Attacks Between Mining Pools," *Financial Cryptography and Data Security*, pp. 63–77, 2015. In BITCOIN'15: The Second Workshop on Bitcoin Research.
7. J. Kelleher. "What Is Bitcoin Mining?" *Forbes*, May 2014.
8. H. Massias, X.S. Avila, and J.J. Quisquater. "Design of a Secure Time Stamping Service with Minimal Trust Requirements,"*20th Symposium on Information Theory in the Benelux*, May 1999.
9. B.P. Eha. "How the World's Richest Nations Are Regulating Bitcoin," *Entrepreneur*, February 2014.

6 Growth of Financial Transaction toward Bitcoin and Blockchain Technology

Chiranji Lal Chowdhary
VIT University

CONTENTS

6.1   OVERVIEW OF THE BLOCKCHAIN

The sole inventor of the cryptocurrency Bitcoin was Satoshi Nakamoto. None seems to have more knowledge of this subject than Satoshi Nakamoto. The blockchain approach is a cryptography-based distributed ledger which empowers trusted trades amid untrusted members in the network. In recent years, blockchain technology has gained its popularity throughout the world, and this technology attracts substantial attention among academicians, investigators, developers, and business experts because of its inimitable trust and security features. On 1 August 2019, the retail business giant Wal-Mart has filed a patent for a stable coin via blockchain which is backed by USD [1].

The common man is worried about security threats that can break and damage cybersecurity, which leads to loss of their personal data and money, as it appears every day in newspapers and social media. Some criminal-minded hackers execute web-based manipulation, thefts, scams, and spying of traditional currency whenever there is an opportunity. To overcome such threats on the financial amphitheater, the chances of breaches in inflated security should be reduced through developing conventional confidential practices. In recent days, the cryptocurrency Bitcoin has gained its worldwide popularity due to the highly secured environment it provides through a blockchain technology that employs diffusion and encryption of information. The blockchain is the backbone of Bitcoin. As there is an increase in the use of Bitcoin and also the need for Bitcoin cybersecurity, every industry wants to adopt blockchain technology to secure their personal data and economic property by difficult-to-trace encrypted transmissions.

In blockchain technology, labeled blocks and chains are used to build a structure for recording the antique data of Bitcoin transactions between different accounts (Figure 6.1). A server controls the whole thing by extensively issuing a hash value for each block of items through time stamping. The blockchain technology is a likely outcome of the ledger technology advanced in addition to dispersed situations. The blockchain technology resolves the problem of multiparty trust in disseminated bookkeeping. The decentralized nature of blockchain technology [2–4]

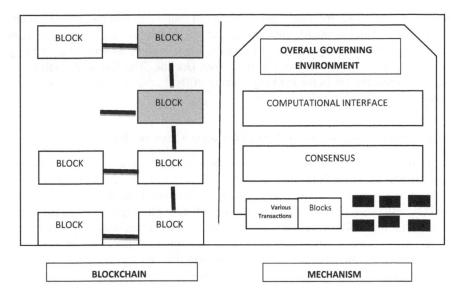

FIGURE 6.1   Architecture of blockchain technology.

makes it tamper-resistant, safe, and reliable to have all embracing consideration subsequent to its start.

Blockchain technology security is still much behind and furthermore, the development in deployment and implementation is required. The risks may arise by attacks from both outside and inside the network. With the increasing popularity of blockchain, the demands are also increasing on protecting the security and privacy of data storage, transmission, and applications. This requires creating new tasks for advancing current security explanations, verification tools, data protection, privacy protection, and information regulation.

This chapter is organized as follows: Section 6.2 describes the astonishing worldwide financial crisis of 2007–2008. Section 6.3 presents the history of Bitcoin and cryptocurrency. Section 6.4 discusses the preliminaries of Bitcoin mining. Section 6.5 is about the aspects of Bitcoin and blockchain technology, which is followed by Section 6.6 where the ways of Bitcoin and blockchain structure of the monetary are discussed. Section 6.7 provides a list of cryptocurrency future scenarios. Section 6.8 covers the impressions of Bitcoin on economy, banking, and finance. Section 6.9 concludes the growth of financial transaction toward Bitcoin and blockchain technology.

6.2   ASTONISHING WORLDWIDE FINANCIAL CRISIS OF 2007–2008

The most devastating financial emergency in the last 10–15 years is the crisis of 2007–2008 [5]. This was the biggest fiscal tragedy that caused immense despair after the Great Depression of 1929–1930. The 2008 crisis started in 2007 by a calamity in the subprime hypothecation bazaar in the United States and advanced to a complete global banking disaster through the failure of the speculation bank Lehman Brothers

on 15 September 2008 [6]. Huge bailouts of fiscal foundations and additional sooth-
ing of financial and economic rules remained working to avert a conceivable ruin
of the biosphere monetary scheme. This predicament stayed, however shadowed by
a worldwide commercial slump, the Excessive Decline. The European obligation
disaster, a predicament in the banking scheme of the European countries by means
of the euro, tracked future [7].

6.2.1 Special Effects of the 2008 Financial Crisis on India

It took time for Indian economists to accept that there are some effects of 2008
fiscal disaster on India. Nonetheless in this period of time Indian administration in
an organization is generally considered a difficult situation in the Indian economy.
The U.S. collapse, which trembled the domain, made slight impressions on India due
to India's robust fundamentals and fewer experience of Indian monetary segment in
the international fiscal marketplace. Possibly, this situation ensured Indian economy
to overcome from this situation speedily. Different from entrepreneurship rubrics of
United States, in India, the marketplace is thoroughly controlled through the regime.

6.2.2 Influence on the Stock Market

In 2008, both the U.S. economy and India's stock market are affected due to the
slowdown of economy. In the fall of 2008, an existing amount of about INR 250,000
crores smeared out in a single-day stock market of India's share market. The Sensex
reduced to 1,000 points in a single day on 10 October 2008 with a previous recapture of
200 points. This enormous drawing happened because India's stock market was mostly
run by Foreign Institutional Investors (FIIs) and participatory notes.

6.2.3 Influence on India's Trade

The trade discrepancy is attainment at disturbing scopes. Due to different cases of
deposition by employees, operative's transmittals, non-resident Indian credits, FII
asset, and others, the present discrepancy is at about ten billion dollars. In case the
interest payment declined and FII invest in the overseas economy increases, India
will come to the same situation that happened in the year 1991, uncertainty about the
foreign exchange capitals reduces and trade discrepancy retains cumulative at the
current rate. Additionally, the foreign exchange assets of the nation have exhausted
through about approximately 55 billion dollars to 255 billion dollars for the month
end of October 2008.

6.2.4 Influence on Exports from India

The United States and some European country reduce their investment on exports
from India. Engineering division, leather, cloth materials, stones and jewels par-
take stayed hit tough as of the fall in the ultimatum in the United States and
European nations. At the same time, India appreciates to trade with United States,
and nearly 15% of trade share done with U.S. in 2006–2007. Indian trading

reduced by approximately 10% in November 2008, at that time the demand also falls in United States and the worldwide demand reduced at the second month consecutively and the demand of cloth monthly trade discrepancy completed ten billion U.S. dollars [8–10].

6.2.5 INFLUENCE ON HANDLOOM SEGMENT, JEWELS EXPORT, AND TOURISM IN INDIA

The Organization for Economic Co-operation and Development (OECD) has changed its policies on the Indian ornaments and jewels manufacturing, handloom, and travel areas, which leads to 50,000–60,000 workers lose their jobs in jewels manufacturing industry that impact also shown in the world wide economy. Additionally, the financial loss at the handloom manufacturing is INR 3,000 crores (at that time in 2007–2008) and the total number of handloom distributes at that time was approximately 5%, which led to excessive job loss in this sector. As the global economy is still experiencing the meltdown, Indian tourism division is seriously affected as the number of travelers from United States and European countries reduced abruptly.

6.2.6 CONVERSATION RATIO DEVALUATION

By the expenditure of FII, Indian currency is roughly less than 20% in contradiction to U.S. currency and hoisted at INR 50 per dollar at an approximate point, forming dread between the shippers.

6.2.7 INDIA'S INFORMATION TECHNOLOGY AND BUSINESS PROCESS OUTSOURCING SECTOR

India's Information Technology and Business Process Outsourcing (IT & BPO) sector was anticipated to attract a total income of 33% or 64 billion U.S. dollars by the end of fiscal year 2008. In the recent past, software service to invest closely two million, an increase of around 375,000 to take software developer in the Information Technology sector. IT segments originated around 75% of their incomes from United States, and IT Enabled Services occupy around 5.5% of entire Indian trade. Consequently, the collapse in the United States certainly influenced the IT segment of India [11,12].

6.2.8 FOREIGN INSTITUTIONAL INVESTORS AND FOREIGN DIRECT INVESTMENT

Generally, when the financial loss impact is on large industry then the Indian national stock market will lead to a loss of a large sum of cash. Certainly, this also has an influence in the business zone, although the minimum cash flow also affects the different Industry sectors. Because of international downturn at that time, FIIs brought back the investments worth about 5 billion dollars; however, the influx of foreign direct investment doubled from about 7 billion dollars in 2007–2008 to about 19.3 billion dollars in the second and third quarters of 2008. This one directed toward the Prodigious Downturn. After that the housing value was cut down by approximately 32%, which is a complete loss for the investor.

6.3 CAUSES OF CRISIS

The initial financial loss in the industry started in the end of the year 2005 and the economy fall started in 2006. This was noticed after housing prices started to go down. In the beginning, realtors admired the situation. The people were understating that the enflamed housing market would return to a further bearable level. The estate agents failed to understand that there are lots of proprietors who hesitated to claim their property. Those days, banks permitted people to obtain home loans up to 100% the property value which strapped banks toward savings in subprime zones.

The real reason for such a situation was the Gramm-Rudman Act. The act permitted banks to engross in trading lucrative products which are sold to stockholders. Such mortgage-backed safeties required home loans as surety. The products formed a voracious request aimed at increasing loans. Hedge funds and few more monetary institutes about the domain possessed the mortgage-backed safeties. The safeties included cutting-edge mutual funds, corporate assets, and pension funds. The banks obligated hacked up the innovative mortgages and further sold their property in tranches. That leads to reduce the value of property.

One reasonable big question was the purchase of unsafe assets by stodgy pension. It was due to insurance company withdrawing the credit limit of the buyers. Such swops were vended by the old-style insurance corporation American International Group (AIG). Once the offshoots mislaid value, AIG failed to consume sufficient cash flow toward all the swops it sold.

One more symbol of the monetarist emergency was in 2007 when banks freaked and appreciated to have to engage the fatalities. Banks were not giving financial support to the buyers respectively. Banks were not ready to accept insignificant debts as warranty. Not a single person purchased a land at that time. This suspicion inside the banking municipal remained the main reason for the 2008 financial crunch [13,14]. The Federal Reserve started driving liquidness hooked on the banking scheme through the Term Auction Facility. That is the only solution present at that time for recovery from the crisis.

Precipitate and Previous History of Bitcoin and Cryptocurrency

At the time of worldwide financial crisis in 2008, Bitcoin played a crucial role in the market to transfer money in the form of virtual currency. In the beginning, cryptocurrencies established a shabby tinge, and equally they stayed mostly linked with black market professions of drug agreements, ransom ware expenses, money laundering, and tax dodging. The cryptocurrency was labeled as the greatest troublesome expertise, often associated with scams or huge Ponzi schemes through the Internet.

Bitcoin success hit news headlines as the value of one unit of the cryptocurrency passed 11,500 USD for the first time. Though this frequently mentioned toward by way of newfangled, Bitcoin consumption happened meanwhile in 2009 and this one was expertly constructed through the genesis block, the first block in the Bitcoin network is the backbone of the network. Actually, if somebody invests 1,000 USD in Bitcoin, and the Bitcoin transaction in that year excessively increases then the individual who spent 1,000 USD gets worth of £36.7 million. People who missed to invest in Bitcoin transaction regretted more at a later stage. Consequently, the transitory antiquity of Bitcoin and cryptocurrency is as follows.

6.3.1 THE STARTING CONCEPTS OF BITCOIN (1998–2009)

Bitcoin was the first recognized cryptocurrency that was attempted by generating online exchanges through ledgers enabled through encryption. Some examples of this are B-money and Bitgold. They were expressed nonetheless completely industrialized.

6.3.2 BITCOIN—A PEER-TO-PEER ELECTRONIC CASH SYSTEM (2008)

The first research article entitled "Bitcoin – A Peer to Peer Electronic Cash System" was authored and mailed to as white paper. This paper was coined by a unanimous person called Satoshi Nakamoto, whose actual individuality is unknown till date. The cryptocurrency user in 2018 is shown in Figure 6.2.

6.3.3 THE BEGINNING OF BITCOIN SOFTWARE (2009)

The first Bitcoin software for mining was released in the year 2009. Bitcoins and dealings were monitored and controlled in the blockchain.

6.3.4 THE FIRST BITCOIN VALUE (2010)

Initially, in the year 2010, Bitcoin was started allocating financial worth for the developing cryptocurrency. Few individuals unequivocally vended the Bitcoin first time in 2010 by exchanging 10,000 of them for two pizzas. Now days, the users purchase a Bitcoin whose value has reached $100 million.

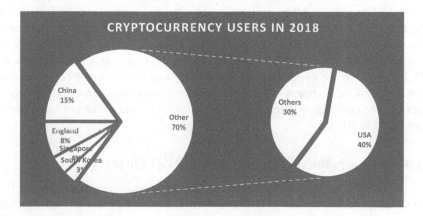

FIGURE 6.2   Cryptocurrency users in 2018.

6.3.5   ARISE OF RIVAL CRYPTOCURRENCIES (2011)

In the year 2011, the Bitcoin is defined as decentralized, distributed and public ledger technology and it was the first cryptocurrency in the world. Afterward, to compete with Bitcoin, altcoin was developed which can give more benefit like transactional processing, security, and giving more incentive towards mining to the miners, in comparison to Bitcoin. Among the first to arise remained Namecoin and Litecoin. In that particular year (2011), the Bitcoin allocated 1,000 cryptocurrencies for new transaction as incentives.

6.3.6   BITCOIN VALUE BANGS (2013)

After the value of one Bitcoin increased to USD 1,000 for the first time, the value rapidly started to weaken. In the year 2013, people who had already invested money in Bitcoin as cryptocurrency had lost plenty of money as the value of Bitcoin had fallen to USD 300, compared to its previous value which stood at two-year USD 1,000.

6.3.7   VARIOUS SCAMS AND THEFTS (2014)

In the year 2014, Bitcoin functions as a medium of exchange. In the same year after the death of the Bitcoin developer, the investors lost 850,000 Bitcoins. The investors were trying a lot to recover those money through different stakeholders of the Bitcoin developer, but they were not able to recover the money. At that time the value of Bitcoin lost is equal to USD 4.4 billion. On prices of 2014 values, those missing coins would be valued USD 4.4 billion.

6.3.8   ETHEREUM AND ICOS (2016)

After the theft of Bitcoins, another cryptocurrency platform originated in 2016 with much expectations, called the "Ethereum." The cryptocurrency used in this platform was called ether which can easily be used by blockchain-based smart agreements and apps. Ethereum's influx was noticeable by the appearance of Initial Coin Offerings (ICOs). An Initial Coin Offering (ICO) is the cryptocurrency industry's equivalent to an Initial Public Offering. ICOs act as a way to raise funds, where a company looking to raise money to create a new coin, app, or service launches an ICO. Interested investors can buy into the offering and receive a new cryptocurrency token issued by the company. This token may have some utility in using the product or service the company is offering, or it may just represent a stake in the company or project [15–18].

6.3.9   A HYPE IN BITCOIN TOUCHES 10,000 USD (2017)

In the year 2017, the digital currency Bitcoin increased dramatically and hit USD 10,000, which was less than $1,000 at the start of that year. Once it reached USD 10,000, the Bitcoin investors started investing more and more money and as a result the cryptocurrency rises to 11 billion USD within a little time and ended with 300 billion USD. Banks such as Barclays, Citi Bank, Deutsche Bank, and BNP Paribas also get benefited due to the rise of Bitcoin. At that time the Fintech manufacturing noticed the underlying concept of Bitcoin that is blockchain.

As Bitcoin and blockchain followed decentralized, distributed, and immutability features, it can be adopted by all centralized authorities to reduce frauds, which is basically done by intruders [19,20].

6.3.10 BITCOIN PAYMENT STARTED BY ONLINE PAYMENT FIRM (2018)

In early 2018, South Korea brought a regulation that needs the Bitcoin dealers to disclose their identities and prohibited the nameless interchange of Bitcoins. At the same time, one online payment firm Stripe proclaimed by April 2018 that it would permit Bitcoin payments to avail its provisions quoting the deteriorating requests, increasing fees, and lengthier business times as per the details on 24 January 2018 [21–23].

6.4 CURRENT STATUS OF BITCOIN (2019)

The capitalization of the digital currency market was nearly 30% greater than before at the beginning of 2019 with a value of over USD 165 billion. Within a month, there was growth in the use of Bitcoin. In March of that year, the Iranian government, which had forbidden cryptocurrencies on its region, agreed to use Bitcoins [24].

The amount of Bitcoin increased by 15% in April 2019. Specialists elucidate that this irregularity is due to the reappearance of major players in the market. The development of Bitcoin happened in the Asian meeting on 2 April 2019, and an unidentified purchaser bought 20,000 Bitcoins, which summed to 94 million USD. Next month, it was noticed that the Bitcoin price sustained to increase. At the end of May 2019, the value of BTC-Bitcoin was equivalent to 8,721 USD. In June 2019, the value of Bitcoin blew all histories. The Bitcoin value reached 10,000 USD very fast, and was continuously increasing [25,26].

6.5 PRELIMINARIES OF BITCOIN MINING

Bitcoin mining is performed by high-powered computers that solve complex computational math problems through Internet. The miner who exchanges or creates new Bitcoins follows a protocol called Proof of Work (PoW) and later the block is placed in blockchain. Mineworkers deliver a community service by safeguarding the net, and the system recompenses for effort.

Once the transaction is broadcast in a system the miners mine the transaction through PoW and solve mathematical puzzles, if the mining is successfully done than the block can be added in the blockchain; this practice can be done in every ten minutes. The apiece innovative block is added to the blockchain; and these dealings are long-established and logged every ten minutes. The mineworkers' competition for the following block starts. For this there is a competition between miners who validate the transaction first. The first to validate transaction can add that block in the blockchain. It follows Longest Chain Rule. All these methodologies of the Bitcoin mining procedure and protocol using cryptographic hash algorithm is mentioned in Bitcoin white paper.

6.6 ASPECTS OF BITCOIN AND BLOCKCHAIN

Bitcoin lost its popularity with the public in 2013 because it was used for drug trafficking in Silkroad, U.S. The price of Bitcoin was rising at the end of 2017 but reduced progressively later in that year. This remained thrilling when used as everyday currency which appears less reasonable. Additional feature of Bitcoin is less sparkling and generating as the eagerness of public online ledgers. The underlying concept of Bitcoin that is blockchain technology gives a hope to international financial organizations to do business through cryptocurrency. It seems after another five years of Bitcoin's appearance, the blockchain will be the ultimate solution to many industries for transactions of their assets in the form of currency.

The entire worth of Bitcoin reached near 300 billion USD at the end of 2017. Basically, Cboe Global Markets, CME Group Inc., and Nasdaq in end of 2017 adopted Bitcoin in their business and were doing more transaction but it is still negligible. Bitcoin was reaching 21 billion USD in 2017 as a cryptocurrency; it started initial coin in 2008. Their fiery development described notices from regulators about the world even beforehand hackers have stolen nearly 500 million USD value of a digital token named NEM from a Japanese cryptocurrency exchange. The massive mainstream of ICOs misplaced considerable worth in the year of 2018. Blockchain was also having a number of hindrances in some high-profile enterprises to drop or place uphold counting a plan by Australia's stock exchange to start by means of the expertise to procedure equity dealings. The firm Wal-Mart Stores Inc. proclaimed in September 2018 that from the starting of 2019, the firm would need dealers for fresh leafy greens to path their goods by means of a digital ledger advanced by International Business Machines Corp., which has consumed deeply on emerging blockchain tools for commercial [27].

In What Way Bitcoin, Blockchain May Possibly Transmute the Monetary Structure

Just imagine that one day, all paychecks are in Bitcoin. Then, what you may feel? Excluding the liquid assets might summersault in worth at somewhat instant in cryptocurrency imminent. Also, it is a must for you to change all cryptocurrency to INR or USD or any other currency to purchase goods at Amazon, pizza at Domino's, or books at Flipkart.

The outpouring in the price of Bitcoin and other digital currency competitors is not matching with routine usefulness. The compensation processor stripe removes Bitcoin sustenance in the start of 2018 by quoting sluggish operation times and more subscriptions. The Stripe product manager Tom Karlo elaborated that "We have seen the desire from our customers to accept Bitcoin decrease."

Now days, Bitcoin has widely spread. It never competed with present financial system like banks and other financial agencies which are encouraging the increase of assets to ten million traders.

Still many people like millennials are not even having a landline that an impending cohort may do deprived of bank accounts so they are happy with protected, peer-to-peer (P2P) cryptocurrency dealings. Blockchain mechanism and the enormous innovative crypto prosperity have unlocked the door to four cryptocurrency commodities, which may attend in a fresh financial command.

6.7   LIST OF CRYPTOCURRENCY FUTURE SCENARIOS

6.7.1   POSSIBILITIES

The Federal Reserve might release itself a digital currency because few global central banks are also discovering. Some leading corporations such as Amazon, Wal-Mart, and Starbucks may come up with their own digital coins which possibly motivate trust and increase widespread recognition. These leading corporations may increase the use of Bitcoin, Ethereum, Hyperledger Fabrics, and other cryptocurrency like Litecoin, FedCoin, which leads to gives more security, reliability, and usefulness to crypto asset value transaction and also in near future many government organizations also used cryptocurrency.

"Virtual currencies might just give existing currencies and monetary policy a run for their money," says International Monetary Fund director Christine Lagarde (2011–2019) in the previous drop. "Citizens may one day prefer virtual currencies since they potentially offer the same cost and convenience as cash – no settlement risks, no clearing delays, no central registration, no intermediary to check accounts and identities," she supposed.

For a long time several tech-related business persons have kept their investment in cryptocurrency to think that in twenty-first century the cryptocurrency price is likely reach to the gold price. In this rush, JPMorgan Chase (JPM) CEO Jamie Dimon says Bitcoin is "a fraud." He and other giants of economics voice assert that profitable banks will endure crucially, cryptocurrencies may stay the fringe, and governments have to save it.

6.7.2   FEDCOIN: AN ALTERNATE CENTRAL BANKS MULL FUTURE OF CRYPTOCURRENCY

Blockchain's possibility to transform the fiscal structure has made central banks to review whether to release their own digital currencies. The prestigious Yale University professors have offered the "FedCoin." In present talk about cryptocurrency, FedCoin may come as a monetary policy which is more supple and powerful. That may even lead to negative interest taxes.

In a situation when a cryptocurrency is represented as a dependable, extensively putative stock of value, persons may break their bonds with banks. People will have some crypto cash in digital wallets along with other liquid assets in mutual funds, stocks, and government bonds.

A reading by Bank of England is ending with a conclusion that a central-bank cryptocurrency may enhance gross domestic product by 3%. The advantages would originate after decrease of "monetary transaction costs that are analogous to distortionary tax rates."

FedCoin is available at the same time when many organizations do not believe in cryptocurrency. Central bank crypto dollars "could endanger the economically and socially important financial intermediation function of commercial banks," cautioned JPMorgan analysts. The influence of insignificant reserve banking to worldwide development, which is rotating apiece of 1 USD of credits hooked on 10 USD in mortgages, could diminish [28,29].

6.7.3   Is This Called as Bitcoin Crash or Cryptocurrency Revolution?

JPMorgan CEO Jamie Dimon puts few points about cryptocurrency in front of the government. They also insist that FedCoin does not support security measures and any body can hack it.

There was big news on 6 February 2018 about Bitcoin which decreased to 6,000 USD on the morning of an important Senate cryptocurrency trial. All the financial regulators were alert at that time. They did not consider any law imposed in cryptocurrency. After few weeks, again Bitcoin price recovered to about 11,000 USD that was previously in the market value of 7,000 USD.

Bitcoin needed to be crumpled before middle of December 2018 by striking the highest overhead of 19,000 USD by impartial as Bitcoin investments started interchange on Cboe Global Markets (CBOE) and CME. The expectation of investments interchange, advertised as authenticated by U.S. regulators, stoked speculations.

6.7.4   Potential System Risks by Bitcoin

The central banker for global central banks, i.e., "The Bank for International Settlements," has cautioned that cryptocurrencies in the future may develop a "threat to financial stability" in case regulators are not attentive to the risk factor in it. The U.S. regulators act to be in performance catch-up. In the month of February 2018, the treasury Secretary Steven Mnuchin conducted a summit on cryptocurrency to describe its functions, regulations on the group of peoples [30].

6.8   IMPRESSIONS OF BITCOIN ON ECONOMY, BANKING, AND FINANCE

Cryptocurrencies remain a troublesome economic invention, which consumes the probable to reform the recent economic assembly and alteration of how banks and financial establishments activate. Bitcoin stands the furthermost widespread arrangement of cryptocurrency that allows digital dealings between two parties deprived of the essence of an intermediate. Each deal is digitally verified in blocks which perform similar to ledgers, and when a block is fully occupied, a newfangled block is formed. Altogether these blocks are linked by means of hashtags linearly in a sequential order called a blockchain. Therefore, each deal is digitally recorded to maintain safety at the highest notch level. However, the dealings are logged, and the info of the gatherings contributing in the conversation is not exposed. The currency can be tracked once it is rehabilitated into money. This community way of handling dealings has shaped the option of an enormous rebellion in the banking subdivision crossways the sphere. The economic control which is fraudulent by the governments and financial organizations supports the use of cryptocurrencies.

6.8.1   Dark Web Power

The dark web is a segment of the web which is not available through the search engine. It is allowed to artificial web which does not fall regularly in the present Internet.

The dark web is reachable only through a distinct software alike Tor Browser, which empowers unidentified penetration of the Internet.

The dark web remains the place where things such as killers, arms, and a lot more unlawful paraphernalia can be found. With use of cryptocurrencies, like Bitcoins, persons can execute unlawful dealings without giving any personal information. Cryptocurrencies like Bitcoins remain a way to authorize such dealings across the world, which will eventually lead to increased cybercrime.

6.8.2 SPECULATIONS

At the beginning of 2015, the value of a Bitcoin remained about 170 USD which reached to more than 15 times in the middle of 2017. There were several ups and downs in the worth of Bitcoins, and such a situation is probable to endure. Because of lots of controversy throughout, still Bitcoin has growing at its speculation. The growing transaction and exchange capabilities of Bitcoins giving huge sold-out in the market as advance.

6.8.3 POLITICIZATION OF CURRENCY

The central banks perform their direct and indirect financial transaction in a centralized manner. Nowadays, by the development of Bitcoins, the situation is altered. The command stayed conferred in the governments, and central banks are instable to the multitudes. This radical change is also modifying the methodologies followed for monetary transaction. To convey safety and permit inspection, central banks and monetary organizations preserve the greatest of all the dealings commenced by the persons. Nowadays through digital currencies, the fiscal supremacy can be confronted by individuals. The leading industries which are independent bodies now also easily deal with Bitcoins. Crypto provides everyone the opportunity to experience clean, sound, and peaceful money. Eventually crypto is accepted on a great scale, while Bitcoins is leading as a safeguard against economy.

6.8.4 APPREHENSION AMONG THE CENTRAL BANKS

There exist some insinuations where Bitcoins are used to clandestinely launder currency to a higher extent. Central banks across the domain have been cautious about Bitcoins as an irrepressible and random system of exchange. Cryptocurrencies are important to reduce the fraud happened in the current bank due to irregularity of giving over draft to the many industries or organization which hamper the Indian economy. So, Crypto and cybersecurity play important role for governments to reduce fraud.

6.8.5 EMERGENCE OF NEW MARKETS

Nowdays, cryptocurrency plays a vital role in the market. Currencies like Bitcoin and Ethereum devise unlocked entrances for an original caring of bazaar which distinct present-day currency bazaar is measured through no one. Cybersecurity will also grow to help bazaar from any kind of fraudulent activities. Now, with the help of

Bitcoin, no transaction fee can be given in comparison to traditional currency transaction. What can be confidently specified is that it is fair the start and the number of potentials are boundless.

6.9   CYBERSECURITY ASPECTS OF BLOCKCHAIN TECHNOLOGY

The blockchain technology is offering an important aspect for cybersecurity. The blockchain provides a platform for a protected network, data, and communications. Blockchain used hash functions and encryption technique to store data as record, which are more secure over various cybersecurity technologies. The existing security system deals with a single-trusted authority to authenticate info or accumulate encrypted data [31]. Due to single-trusted authority, many attackers used various attack like SQL injection, denial of service attack and stolen the information easily. Blockchains deals with current security issues by creating a decentralized platform, and they do not need trust of a distinct associate of the group or network or the authority. Blockchain is a trustworthy system where all the data are stored in a distributed public ledger across the nodes with a smart contract which is immutable in nature. Most members of a group agreed to the same information that will be able to secure the group much better than a group made up of one leader and a host of members relying on the leader for their information, particularly when bad actors might come in the form of group members or even as the leaders themselves.

The blockchain technology is used to enhance the cybersecurity of things on the Internet, data storage and sharing, network security, private user data, World Wide Web navigation and usefulness. Internet-of-Things (IoT) has the foremost private blockchains to enable device admission control in the network to secure track data management and prevent malicious access. In addition, blockchain is used to advance software organization protection over P2P upgrade propagation to provide recognition, encryption, and seamless secure Internet computer data transfer.

Blockchain application is used for securing practical IoT information, terminology, and distinguishing malicious results. The blockchain protocol is between the network's software and transport layers that use Bitcoin-like token incentives and that extravagance as voting power components (Figures 6.3 and 6.4).

Accustomed data storage and exchange area unit of public and private shared ledgers area unit remove one failure supply at intervals of a given storage system, protecting the information from a change of state. That is, blockchain helps ensure this data stored in the cloud remains resistant to unauthorized changes, hash lists provide a quest for details that will be preserved and kept securely, and updated information will be checked as identical from delivery to receipt. During a shell, blockchain enhances the storing and sharing of information by building a decentralized network which uses client-side encoding in which information homeowners can handle their information in total traceability.

Most system security identification works use blockchains to boost Software-Defined Networks (SDNs) and use owners to decentralize and actively store basic information for verification. Blockchain-enabled SDN controller technology using a group structure is used in these works. For P2P communications between hubs in the network and SDN controllers, the design uses public and private blockchains to render the blockchain suitable for security arrangements [32].

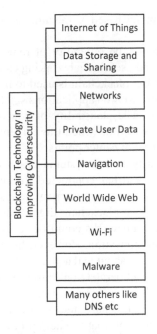

FIGURE 6.3   Blockchain technology is applicable to improve cybersecurity.

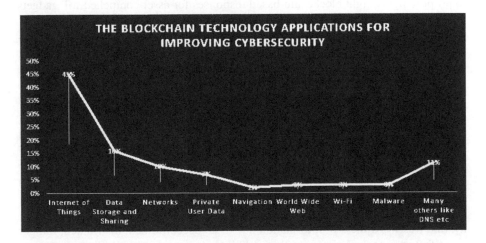

FIGURE 6.4   Blockchain technology applications for improving cybersecurity.

The private information of the customer is basically not put in the blockchain due to its distributed features. The explanation could be due to the blockchain's idea of irreversibility (everyone has a record duplicate), which makes it difficult to use for security purposes, particularly in information insurance. Common client gadget inclinations are encoded and placed on the blockchain in current approaches to be recovered by that client uniquely. We also explore comparisons between blockchain proof of work (PoW) and proof of validity agreement schemes, where hubs are given a rating to assess their legitimacy based on the number of hubs-confined connections with others.

Blockchain is used to boost the reliability of remote Internet parts relevant to World Wide Web research and usefulness by taking care of and testing local record passage control data. Likewise, blockchain is used to help investigate the right page through correct DNS records, use web applications safely, and talk to others through encrypted, encoded systems. The plausibility of consortium blockchain was used to execute these courses of action in which the agreement method is forced in the framework by a preselected set of center points.

Blockchain for IoT security in IoT systems has been asserted as a squeezing requirement of the business and has contracted the most extreme need aimed at development and requirement, in spite of momentum research demonstrates the way that pretty much each artifact on blockchain digital security in the writing focuses out that the security of IoT frameworks could be revived on the off chance that it is bolstered with blockchain innovation. However, little is known and talked about variables identified with choices about and achievability to embrace this innovation, and how and where it can be deliberately put into use in an unmistakable setting to correct current IoT security hazards/dangers, taking into account the creative mind and future vector production in this particular space after that. In this way, building up some quantifiable rules and tools that can help fill this out clearly in the writing is important for future research. In addition, it could be another area of further research to propose lightweight blockchain-based responses for asset-compelled IoT gadgets (running on the edge of the system).

In present day, blockchain used for Artificial Intelligence (AI) information security where the data are collected from different devices like sensors, actuators which is generally used in Internet of Things (IoT) devices. Man-made brainpower (AI) and its subordinates have been utilized as integral assets to break down and process the caught information to accomplish successful thinking intending to security issues.

In spite of the fact that AI is ground-breaking and can be locked in with dispersed processing, a beguiling investigation would be created when undermined or exploitative information is deliberately or accidentally coordinated by a malevolent outsider dependent on ill-disposed sources of info. Blockchain as a famous record innovation can possibly be utilized in various regions of the Internet.

The adoption of blockchain technology guarantee gives legitimacy, dependability, and trustworthiness of information due to its decentralized, distributed, tamper proof features. At the point when the credibility and unwavering quality of information can be guaranteed, increasingly secure and reliable results can be created by AI. A future research bearing could be the investigation of blockchain for the security of AI information in Business to Business (B2B) and Machine to Machine (M2M) conditions.

Sidechain security technology has to minimize the difficulties (essentially execution) associated with primary blockchains most of late rising as a separate chain linked to the fundamental chain, in parallel with exchanges. We envision a transmitted multi-blockchain world sooner rather than later, in which distinctive fundamental chains and sidechains function together in different situations.

6.10 BLOCKCHAIN AND BITCOIN CYBERSECURITY RISKS

6.10.1 Is Blockchain Fail-Safe?

True, the randomness of the data transactions within the context of blockchain and their strong encryption means that using malware or other exploits, neither the blocks nor the chain can be duplicated or infiltrated.

According to Lexology, however, if the data at one end of the chain is false or inaccurate, the other party has no way of correcting it – unlike wire transfers and card transactions that can be halted indefinitely if misconduct is identified quickly.

In a nutshell, white-collar criminals could take advantage of the excitement of their peers over blockchain to commit completely secure acts of fraud that could not be discovered for days, making it difficult to maintain Bitcoin cybersecurity.

All human interactions, including business deals, are based on trust. Blockchain businesses are justified in implementing a system that already proves to be a game-changer and has the potential to be even more effective. But if business leaders over-zealously use blockchain and do not carry out due diligence with potential customers or partners, they are open to exploitation.

While it is not easy to project the development of blockchain, there is little doubt that it will remain part of the larger technology landscape.

6.10.2 Cybersecurity Plays a Key Role in Today's Market

One of the most important positions in the modern business world is the role of the cybersecurity specialist, particularly if you are interested in innovative technologies such as Bitcoin and blockchain.

If you want to follow a high-tech career and continuously change the cybersecurity sector, pursuing a cybersecurity master's degree can be the first step toward an impactful career.

Explore the online Master of Science in cybersecurity program at Maryville University to see how it can prepare you for success in working with Bitcoin, blockchain, and other innovative developments in the cybersecurity sector, which is relevant and often financially rewarding.

CONCLUSION

The online bazaar is enthusiastically emerging in these years. The highest online cryptocurrency is Bitcoin. The worth of dealings settled by the contribution of Bitcoin upsurges every year. Conversely, an authorized financial bazaar has not been formed aimed at cryptocurrency. On behalf of cryptocurrency, there wasn't any creation of financial institution, which organizes the procedure of raising the financial base. There wasn't any form of a central bank for Bitcoin; therefore, there is not at all any fiscal rule for Bitcoin. However, few commercial banks remain commencement to identify such money. Few savings funds take capitalizing in Bitcoin for some years and attaining extraordinary effectiveness of asset.

So the main questions rising here are as follows: Is it possible for Bitcoin to be the central money for financial businesses and payments resolved through the Internet in the upcoming years? This may be possible in case this is substituted authorization currency part through the choice of centralized banks. But still, it will take year to come up with such a challenging decision. Nowadays, maximum of the businesses are run through online purchases.

Those who are working with cryptocurrency across the borders, by purchasing and vending goods and services, may be gaining a huge profit. In cryptocurrency and blockchain technology, there is no transaction fees required for any transaction. Cryptocurrency transactions stay near rapid. The transmission supports directly which is taking merely ten minutes for overheads to be authenticated in the blockchain. The financial exchange taking place by a Bitcoin payment is absolutely secure. If the users have enough money in his wallet then the transaction must successful. The cryptocurrency is just a P2P financial scheme, so there are no taxes at all. It can be said that there are no government taxes or any deal charges. The blockchains formed after cryptocurrency dealings to make a clear, safe account, which can be followed and confirmed. A drawback of this cryptocurrency technology is the undefined future. At present, even a single Bitcoin is so costly, but how long this may continue. Also, Bitcoin isn't legal in various countries so this is not an internationally accepted currency.

REFERENCES

1. Bayern, S. Of Bitcoins, independently wealthy software, and the zero-member LLC. *Northwestern University Law Review, 108*, 1485, 2013.
2. Giungato, P., Rana, R., Tarabella, A., & Tricase, C. Current trends in sustainability of Bitcoins and related blockchain technology. *Sustainability, 9*(12), 2214, 2017.
3. Hileman, G., & Rauchs, M. Global cryptocurrency benchmarking study. *Cambridge Centre for Alternative Finance, 33*, 36–45, 2017.
4. Narayanan, A., Bonneau, J., Felten, E., Miller, A., & Goldfeder, S. *Bitcoin and cryptocurrency technologies: A comprehensive introduction.* Princeton University Press, St. Princeton, NJ, 2016.
5. Bouri, E., Roubaud, D., & Shahzad, S. J. H. Do Bitcoin and other cryptocurrencies jump together? *The Quarterly Review of Economics and Finance*, 2019. https://doi.org/10.1016/j.qref.2019.09.003.
6. Bouri, E., Gupta, R., & Roubaud, D. Herding behaviour in cryptocurrencies. *Finance Research Letters, 29*, 216–221, 2019.
7. Brauneis, A., & Mestel, R. Price discovery of cryptocurrencies: Bitcoin and beyond. *Economics Letters, 165*, 58–61, 2018.
8. Fry, J. Booms, busts and heavy-tails: The story of Bitcoin and cryptocurrency markets? *Economics Letters, 171*, 225–229, 2018.
9. Yuan, Y., & Wang, F. Y. Blockchain and cryptocurrencies: Model, techniques, and applications. *IEEE Transactions on Systems, Man, and Cybernetics: Systems, 48*(9), 1421–1428, 2018.
10. Corbet, S., Meegan, A., Larkin, C., Lucey, B., & Yarovaya, L. "Exploring the dynamic relationships between cryptocurrencies and other financial assets." *Economics Letters, 165*, 28–34, 2018.
11. Dierksmeier, C., & Seele, P. Cryptocurrencies and business ethics. *Journal of Business Ethics, 152*(1), 1–14, 2018.
12. Temin, P. "The Great Recession & the Great Depression." *Daedalus, 139*(4), 115–124, 2010. Financial Crisis of 2007–2008. https: //en.wikipedia.org/wiki/Financial_crisis_of_2007%E2%80%932008

13. Agree, T. T. E. Worst financial crisis since great depression. *Risks Increase if Right Steps Are Not Taken, Business Wire News Database.* http://ca.news.finance.yahoo. com/s/13022009/34/biz-f-business-wire-three-top-economists-agree-2009-worst-financial-crisis.html. 2009.

14. Eigner, P., & Umlauft, T. S. The Great Depression (s) of 1929–1933 and 2007–2009? Parallels, Differences and Policy Lessons. *Parallels, Differences and Policy Lessons (July 1, 2015). Hungarian Academy of Science MTA-ELTE Crisis History Working Paper,* (2), 2015.

15. Eichengreen, B., & O'Rourke, K. A tale of two depressions: What do the new data tell us?VoxEU.org, *8,* 2010. https://www.advisorperspectives.com/newsletters10/ pdfs/A_Tale_of_Two_Depressions-What_do_the_New_Data_Tell_Us.pdf

16. Temin, P. The great recession & the great depression. *Daedalus, 139*(4), 115–124, 2010.

17. Williams, M. *Uncontrolled risk: Lessons of Lehman brothers and how systemic risk can still bring down the world financial system.* McGraw Hill Professional, New York, NY, 2010.

18. Marr, B. A. Short history of Bitcoin and crypto currency everyone should read. *Dostupno na* https://www.forbes.com/sites/bernardmarr/2017/12/06/a-short-history-ofbitcoin-and-crypto-currency-everyone-should-read, 2017.

19. Sutradhar,D.USfinancialcrisis:CausesanditsimpactonIndia.pp.1–9.https://www.scribd. com/document/127532103/US-Financial-Crisis-Causes-and-Its-Impact-on-India.

20. Mukeshchandra, P. C., Charles, V., & Mishra, C. S. Evaluating the performance of Indian banking sector using DEA during post-reform and global financial crisis. *Journal of Business Economics and Management, 17*(1), 156–172, 2016.

21. Bardhan, A. *Of subprimes and subsidies: The political economy of the financial crisis.* University of California, Berkeley, 2008.

22. Kolb, R. W. (Ed.). *Lessons from the financial crisis: Causes, consequences, and our economic future* (Vol. 12). John Wiley & Sons, Hoboken, NJ, 2010.

23. Chitale, R. Seven triggers of the US financial crisis. *Economic and Political Weekly,* November 1, 20–24, 2008.

24. Nachane, D. M. The fate of India unincorporated. *Economic and Political Weekly, 44*(13), 115–122, 2009.

25. Varghese, A., & Salim, M. H. Handloom industry in Kerala: A study of the problems and challenges. *International Journal of Management and Social Science Research Review, 1*(14), 347–353, 2015.

26. Murphy, R. T. Asia and the meltdown of American finance. *Economic and Political Weekly, 43,* 25–30, 2008.

27. Aalbers, M. B. The great moderation, the great excess and the global housing crisis. *International Journal of Housing Policy, 15*(1), 43–60, 2015.

28. Sodhi, N. S., Bickford, D., Diesmos, A. C., Lee, T. M., Koh, L. P., Brook, B. W., & Bradshaw, C. J. Measuring the meltdown: drivers of global amphibian extinction and decline. *PloS one, 3*(2), e1636, 2008.

29. Lewis, P. H. *The agony of Argentine capitalism: From Menem to the Kirchners.* Praeger, Westport, CT, 2009.

30. Davis, E. P., & Karim, D. Could early warning systems have helped to predict the sub-prime crisis? *National Institute Economic Review, 206*(1), 35–47, 2008.

31. Sivaraman, B. The impact of the US meltdown on the Indian economy. 1–7, 2008. https://www.scribd.com/document/45480775/The-Impact-of-the-US-Meltdown-on-the-Indian-Economy.

32. Kharif, O., & Leising, M. Bitcoin and blockchain. Bloomberg QuickTake, 2018. https:// www.bloomberg.com/quicktake/bitcoins.

7 A Brief Overview of Blockchain Algorithm and Its Impact upon Cloud-Connected Environment

Subhasish Mohapatra
ADAMAS University

Smita Parija
C.V Raman Global University

CONTENTS

7.1   INTRODUCTION

The financial services industry plays a vital role in commodity transaction and society welfare since it allows savings for future and investments. It provides protection from risks and safeguards business enterprises [1]. Information technology has changed the basis of the industry over decades. It enables an extensive increase in transactions and diversification of products [2]. However, the current pace of Information and communication technology (ICT) innovation in business sector has a great potential for blockchain. This mostly happens to reduce security burden by decentralizing it, i.e., lodged within the industry. The comprehensive view of blockchain technology serves as a paradigm shift in security and innovation [3].

The success story of Bitcoin adds a new page in blockchain technology. This chapter provides a comprehensive overview of blockchain algorithm in numerous fields including healthcare service, financial service, cloud, IoT, etc. There are still many provocations in blockchain technology in terms of scalability and security. That needs to be justified by blockchain algorithm to achieve immutability in network. The confront of blockchain technology has reached on zenith recently. Blockchain ledger is nothing but a decentralized cryptic data storage. Bitcoin is the most successful prodigy of blockchain, which was introduced in 2008 and implemented in 2009. Here, all fully committed transactions are stored in a list of blocks. The chain of records grows as new blocks join subsequently. It is based on cryptography and distributed consensus algorithm. It can further enhance user security and ledger consistency. Blockchain is inculcated with the trait of persistency, auditability, and decentralization. It ascent a new opportunity of efficiency and cost saving. It can provide right framing for digital asset, online payment, and transfer of remittance. In addition to this, it can check upon black money laundering. It can also be favored by many sectors, e.g., service industry, cloud, finance, and IoT. Enterprise that can use blockchain technology can gain the faith of its consumers. So ultimately, blockchain has the potential to raise the share market value of that company [4].

This chapter contributes the following: (1) cloud settlement and (2) blockchain compliance. Embedding cloud computing in blockchain carried its further avenues for research.

7.2   OVERVIEW OF BLOCKCHAIN ALGORITHM

Algorithms have the capacity to govern the activities of humans and give insightful knowledge of the data that is considered substantially logical. Rather than discussing the everyday use of algorithms in socio-technical-economical systems, we explain how to administer it more precisely in blockchain technology, as well as discuss about the significance of studying algorithm in process implementation that impacts on whole innovation.

Block chain ecosystem [5, 6] is an open-source, decentralized system appeared differently in relation to the authority of brought together and corporate programming frameworks, in this manner touching off significant change. Pressures exist between individuals in the digital currency network who might want to coordinate cryptographic money into organizations and those that consider digital currency to be an extreme utilization of algorithmic power, liberated from institutional impact.

Blockchain has become a buzzword in academic research and industry. Few companies have gained a huge success by the use of blockchain recently. Blockchain transaction can be made without any intermediary control. Most of the industries are marching toward blockchain because of its immutability [1, 2]. The next-generation Internet system is springing up toward blockchain technology, so we need a more robust algorithm. Recent advancements in algorithm can solve the issues related to blockchain size and security. We can design such an algorithm that protects against privacy leakage and high-frequency trading. As large block sizes need larger storage space, ultimately the propagation in network will be slow. This will gradually make the system control more centralized. Unhurriedly, blockchain miners hide their blocks for more revenue in future. This chapter contributes a technical report on blockchain algorithm and impact of cloud in blockchain design principle. The chronological shift in algorithm focuses upon gradual research in blockchain.

Blockchain formulators are operating on persistency in transaction. It ensures that each sole originator controls their algorithms in its own term of personal data related to identity. To manage the growing list of records, we need a transparent algorithm that also provides privacy and security. A good version of algorithm can provide complete security and verifiability over a connected infrastructure. The algorithm designed for blockchain environment targets upon some specific objectives. It is an agreement between the nodes that participate in entire blockchain network. The algorithm must fulfill some common objectives that drive the program in global scale. To carry out synchronized ledger transaction, we as researchers hope that a good blockchain algorithm must transform business world and financial transaction into a better form. In this chapter, we provide a brief overview of the chronological advancement of algorithm in blockchain environment, but we still need an opulent algorithm that can lift certain crucial aspects such as speed, security, and integration. Blockchain algorithm can execute transaction with decentralized ownership. It is a complete solution for trade, retail, banking, and hospital sectors, where we need an intermediary loyalty program to streamline access information. To implement good business logic over blockchain, every organization needs a complete transparent algorithm that can verify lineage of data, hyperledger fabric, and peer to peer access. A good algorithm will make a complex transaction more efficient and less expensive. For example, amazon quantum ledger database (QLDB) algorithm provides right solutions for each job over cloud. The algorithm used in QLDB can keep track of usage metric of each node in terms of memory, storage resource, and end-to-end blockchain platform [7–9].

In blockchain, the tasks of algorithm are as follows:

- Verifies signatures of content.
- Confirms balances in transaction.
- Decides validity of a block.
- Determines how miners validate a block.
- Establishes a pattern how a block moves in blockchain.
- Provides a procedure for creating new coins.
- Tells the system how to determine consensus.

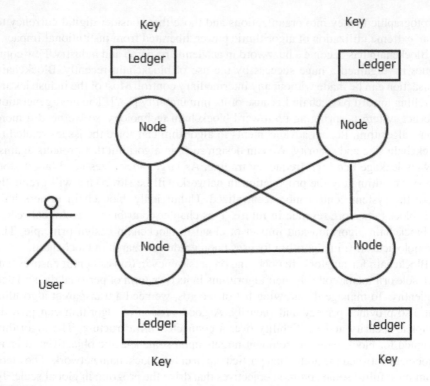

FIGURE 7.1 A blockchain node interaction and overview of ledger.

Algorithms help miners in monitoring and validating a block. They build a logic flow just like protocols, and the fundamental instructions are built up by blockchain algorithm. The autonomy in blockchain infrastructure is a desired outcome. To initiate a transaction, we need to determine which block is going to make its path into the chain, and it provides a consensus to check which chain is correct. Algorithms functionally describe an underlying protocol to achieve desired goals.

Bitcoin and Ethereum are the emerging protocols in blockchain application. They focus upon establishing the pragmatic rules to set up the "engines" and, at the same time, determine who does what action and how. As end users, we then implement the pseudocode of algorithms to make transaction with coins. It will execute a smart contract [10, 11] and give boost to create innovative business models. The algorithms are giving a high-end up-to-date setup that makes the protocols more useful.

There is a distinction between the state and the action. We cannot change the rules, but we can create a series of actions, instructions, and processes specific to our business models.

7.3 TYPES OF ALGORITHMS IN BLOCKCHAIN

After the meaningful implementation of blockchain and the Bitcoin cryptocurrency in 2009 by Satoshi Nakamoto, many algorithms have been highlighted. Development of algorithms is continuously advancing, which also aims to solve the interoperability challenges of existing blockchain infrastructure like the proof-of-work (PoW) [12–14] algorithm system. PoW and Proof of Stake (PoS) are facilitating *consensus algorithms*. They collaborate and provide all blockchain nodes to potentially agree upon blockchain rule and also check on double spending, i.e., if an intruder attempts to spend the same coins more than once, it violates the ethical policies of blockchain algorithms.

7.3.1 CONSENSUS ALGORITHMS

Consensus algorithms are helpful in buying coins or running a node. The main intention of designing consensus algorithms is to accomplish reliability involving multiple nodes on networks, to ensure all nodes adhere to the guidelines. Nodes characterize consensus in Bitcoin, not miners. Chain is associated with consensus along with the most work. If you fork and change the PoW, you will not have the mining capacity to verify it. Nodes acknowledge, validate, and also replicate the transactions, and at the same time validate, replicate the blocks, as well as serve and store the blockchain. Nodes even characterize the PoW algorithm that miners have to employ. Nodes keep up the protocol, not miners. The presentation of blockchain brought the advancement of consensus algorithms, and many more algorithms have been framed which are expected to fix the constraints of the first ever PoW algorithm system.

7.3.1.1 Consensus Algorithm

A consensus block generation algorithm is used to establish an agreement through which a blockchain network gains consensus. Public (decentralized) [15, 16] blockchains stimulate entire distributed systems, and since they are not relying upon any central authority, the distributed blocks need to agree on the validity of transactions. The existence of consensus algorithms emerges into picture. They assure that the protocol rules are followed and guarantee that all transactions achieve reliability; hence, double spending on coins will be checked. Before we discuss in depth about different flavors of consensus algorithms, it is vital to conceptualize the differences between an algorithm and a protocol.

7.3.1.2 Consensus Algorithm Versus Protocol

The terms "algorithm" and "protocol" are considered as pillars to any emerging application, but their pragmatic views are not the same. As a conventional term, authors define a protocol as the abstract rules of a blockchain, and the algorithm stands upon the mechanism through which these rules are being structured. Besides being massively used on financial transaction, the blockchain technology can penetrate into every corner of businesses and would be capable of numerous business use cases. But regardless of the context, a blockchain network is giving a primary focus to a protocol that can resolve how the system is supposed to work, so all the different parts of the system and all participants of the network will need to follow the rules of the underlying protocol.

While the protocol regulates what the procedures are, the algorithm tells the system what steps to take in order to comply with these rules and produce desired results. For instance, the consensus algorithm of a blockchain determines the validity of transactions and blocks. So, Bitcoin and Ethereum are protocols, while PoW and PoS are their consensus algorithms.

To further illustrate, consider that the protocol set a standard of jurisdiction for nodes, i.e., how blocks should interact, how the data should be transmitted between them, and what are the feasible logic for a successful block validation. On the other hand, the consensus algorithm is accountable for validating the balances and signatures, authenticating transactions, and for literally executing the validation of blocks, and this entire phenomenon relies on network consensus.

7.3.2 DIFFERENT TYPES OF CONSENSUS ALGORITHMS

There are several types of consensus algorithms. The most common implementations are PoW and PoS. Each one has its own advantages and disadvantages when trying to balance security with functionality and scalability.

7.3.2.1 Proof of Work (PoW)

PoW architecture use the concept of consensus algorithm. It is widely used by Bitcoin and many other cryptocurrencies. The PoW algorithm is a vital part of mining process.

The objective behind PoW is to guaranty numerous hashing procedures, so more computational power consumption declares more trials per second it need, and subsequently, miners who possess high hash rates have a potential chance to select a full proof solution for the next block (a.k.a. block hash). Enhancement of PoW consensus algorithm depends upon selection of miners and procedure of validation for a nascent block of transactions. It adds this block to the blockchain if the distributed nodes of the network reach consensus and agree that the block hash provided by the miner is a valid PoW.

7.3.2.2 Proof of Stake (PoS)

The PoS consensus algorithm came to reality in 2011 as an alternative solution to PoW. Although PoS and PoW share a common objective, they still deliver some fundamental discreteness and eccentricity, generally at the time of validation of new blocks.

The identity of PoS consensus algorithm replaces the PoW mining strategy with a clear-cut procedure where blocks need validation according to the stake of the participants. The validator of each block (also called "forger" or "minter") is decided by an investment of the cryptocurrency itself, not by the amount of computational power allocated. Each PoS system implements the algorithm in various ways, but in general, the blockchain is shielded by a pseudo-random election process that speculation of the node's wealth determines the coins' age (how long the coins are being locked or staked) – along with a randomization factor.

The Ethereum blockchain is currently centered on a PoW algorithm, but the Casper protocol will shift the network infrastructure from PoW to PoS. It is an attempt to boost the network's scalability [17–19].

As previously mentioned, the consensus algorithms are crucial for maintaining the integrity and security of a cryptocurrency network. They are delivering a potential cause for distributed nodes. It is reaching consensus by which version of the blockchain is the potential deciding factor. Agreeing upon the current blockchain nature and state, it is vital for a digital economic system to work in cohesive infrastructure properly.

The PoW consensus algorithm has emerged as one of the best solutions to the Byzantine Generals Problem, which affirms the creation of Bitcoin as a Byzantine Fault Tolerant system. This means that the Bitcoin blockchain is highly resistant to attacks not only because of the decentralized network but also because of the PoW algorithm; hence, the majority of attacks are fenced by this technology. The high costs involved in the process of mining make it very baffling and unlikely that miners can invest their resources to dismantle the infrastructure.

7.4 MINING ALGORITHMS

The components of data mining are of three types: *Clustering or Classification, Association Rules, and Sequence Analysis.*

Clustering/classification is a technology in which a grouping rule is generated from a set of odd data. The rule will be used to classify future data. An association rule is a well-known approach where it implies a precise association among a set of objects in a database. Sequence analysis is used to establish patterns that present in a chain of sequence. A variety of algorithms are propounded to implement such aspects of data mining. The foundation of blockchain mining algorithm explains how it works. In this section, a brief explanation is given on how mining works, and it is composed of four steps as follows:

Step 1. First a user signs off on a transaction from the application wallet and sends certain cryptic token.

Step 2. The transaction is broadcasted by the application wallet and waits for a miner. The pending transaction is termed as "unconfirmed" according to blockchain. If it is not plucked by a miner, then a local data structure pool is maintained for this unconfirmed transaction.

Step 3. Network miners depicted as nodes select the transaction from the local pool of transactions, and a block is formed with some metadata information. Every miner assembles their own block of transactions.

Step 4. Transactions are taken and assembled to a block. Miners create a block of transactions. So, all the nodes present in blockchain need to register with a transaction. So, validation of the block is necessary before initiating the transaction. The foremost activity of the block is to get signature of proof, i.e., PoW. The signature is a combination of very complex math challenges. This is uniquely maintained for each block. This process is significantly termed as "mining," e.g. a hash function is popularly used by miners in blockchain. A hash function picks up an input as a string of numbers and letters, and subsequently, it returns into a 32-digit string which is a combination of random letters and numbers (hash output).

7.5 TRACEABILITY CHAIN ALGORITHMS

The entire gamut of Traceability is designed to provide authenticity to a digital infrastructure at its nascent stage of development. It focuses upon transparent security practices for any electronic transaction. It would gather additional information to enhance internal node performances. Subsequently, it outlines drafting activity of each single node present in a supply chain model. Blockchain is deciding framework for large chunk of data in commercial transaction. Here massive data is propagating across every single node. The idea behind traceability chain algorithm is depicted as follows: it provides a unique identifier for a traced product. Not only this algorithm provides trace for off-chain data, but it also ensures consistency and data restriction for high-dimensional data. The mother of traceability chain algorithm is a distributed computing infrastructure. The sole objective behind traceability chain algorithm is to forecast traceability decisions swiftly. As a consequence, this emerging technique checks against irrelevant data problems and enhances traceability in blockchain. Nowadays, traceability chain algorithm provides high-end solution for diamond transaction, intellectual property, and pharmaceutical industries. Some recent developments of traceability application in linear blockchain mining algorithm are IBM crypto anchors and Oracle smart contracts. Though the traceability chain algorithm runs faster than a consensus algorithm, each traced product needs a potential inference mechanism. This is the smart solution against data theft. Recently, numerous business houses are facing problem to trace a product precisely. As they do not have previous correct information in track, it is corresponding to repetitive node in blockchain ecosystem. With blockchain, it is possible to constantly trace the history of a transaction. A novelty approach called **Takagi–Sugeno Fuzzy cognitive maps** concentrates on the traceability chain algorithm. The consistent analysis of traceability algorithm in blockchain mining is discussed in this section. It is constantly reshaping objective functions for optimized decision computations among stakeholder nodes by a constraint method. Thus, this traceability improve product visibility succeeds by reducing mining efforts while the traceability sequence is organized.

Blockchain traceability deals with improving the coordination of the supply chain ecosystem; it gives more precise information about the spatial aspect and state of transactions so that a node can exploit this potential information as inputs for its planning activity. Traceability solution imparts a strict sway of transactional lifecycle, being able to trace every digital asset moved in electronic transaction in blockchain. To originate a complete traceability infrastructure, there must be a sincere effort to track movement of activities on the blockchain; blockchain is delivering each item that links in the transaction a virtual identity. Therefore, objects need an association with sensors or a tag that can store information about items and transfer them to the platform, for example, QR codes, RFID, or wireless sensor networks. Traceability chain algorithm is the backbone of a distributed database of records that uses blockchain for exchanging digital assets. The novelty approach behind this set of algorithm provides hierarchical cognitive learning. The traceability chain algorithm provides optimum traceability to blockchain. Machine learning plays a major role to forecast tracing on high-dimensional data set. This set of algorithm focuses upon fuzzy inference mechanism that can predict trace of block. Traceability detection algorithm in blockchain is an indispensable system to trace record accurately. It can prevent data tampering of sensitive information. It adopts an emerging technique to deal with data explosion problem. It adopts a dynamic control of blocks in block-chain model. It effectively considers user demand to relieve data explosion problem. It exhibits a bright future for supply chain management. Traceability chain working mechanism is as follows:

1. Event data learning technique is adopted by considering past data and nature of data.
2. Forecast the pattern of nodes over blockchain model.
3. Precisely capture traceability information.
4. Upload cryptic key information to blockchain.

This emerging technique is still under research.

7.6   BYZANTINE FAULT TOLERANCE (BFT)

The **Byzantine Fault Tolerance** (BFT) is designed to estimate the reliableness of uniformly segregated computing systems. In this process it resolves on components breakdown mechanism and result of crashing information is being mitigated by this algorithm. Conquering BFT is one of the most pivotal challenges directed by block-chain technology. BFT inscribes a mechanism in which two nodes would exchange information safely across a digital infrastructure, making sure that they are showcasing the same data.

The vital effect of "**Byzantine failure**", it is focussing a component that is a server can incoherently give breakdown information by delivering numerous symptoms in each period of time. It is the scenario in which active users agree on a strategy to avoid complete system failure. BFT [20–22] is also projected as source congruency, error avalanche, Byzantine Generals Problem, and Byzantine agreement problem.

Byzantine fault tolerance

- Nodes fail arbitrarily
  - they lie
  - they collude
- Causes
  - Malicious attacks
  - Software errors
- Seminal work is PBFT
  - Practical Byzantine Fault Tolerance, M. Castro and B. Liskov, SOSP 1999

BFT is conceptualized as a logical conundrum that depicts communication problem when they are dwelling upon a common set of problems. The beauty of BFT is, it can continue its operation even if certain nodes behave in an abrupt way.

The widespread application of BFT in blockchain is an ongoing effort. It can tolerate upon a class of failures so that each class can work in an autonomous way. It does not impart any restriction among peers in blockchain. This tremendous application is used in nuclear plants, space, aviation industry, etc.

7.7 CHOOSING THE RIGHT ALGORITHM

To build an emerging digital infrastructure, we need blockchain as a service that comprises numerous algorithms to deal with complex back-end ledgers. It can assist any business model to decentralize its application over cloud. It is a challenging task for any business owner to choose right algorithm for distribution of vital information over web. A right algorithm can assist blockchain developer to integrate information over all possible circumstances. This is the reason we need a smart blockchain algorithm that can deal with all possible circumstances. An algorithm can assist in framework building. It delivers clear-cut information about peer node, master node, and tracking history in blockchain. In a digital ecosystem, study of peer nodes' behaviors is vital before applying blockchain technology. It can further provide immunity against malicious actions. The entire perspective of an algorithm is not only to provide automatic control to nodes but also to judge scalability of business domain. An algorithm in blockchain suggests an innovative structure in terms of transaction tracking and security. The various algorithms and their features are explained in this section. The effect of decentralization and breach of security to remote nodes can be captured by choosing right algorithm. For creating a smart economy, the authors suggest some new aspects of tracking events over cloud-connected infrastructure. It can secure digital identity to create smart contracts among users. Recently, IBM launched a hyperledger fabric framework. It is based on blockchain channel partitioning. For example, NEM is a platform-specific blockchain technology developed in Java. This provides solutions to payment service, messaging, and asset authentication. At the time of blockchain infrastructure development, choosing a right algorithm can provide an instance to keep technology in right track. Selection of a right algorithm is a fascinating solution in blockchain.

7.7.1 BLOCKCHAIN (NON–PROOF OF WORK)

When peers make transactions in a consent infrastructure (networks where all peers are known and trusted), blockchain algorithm exploits the potential of high-end mathematical operations, which can be replaced with simpler but less secure alternatives. Non-PoW blockchain algorithms generally select a leader. As a result, non-PoW blockchain is susceptible to Denial-of-Service (DoS) attacks, but it can showcase a better alternative for democratized transactions.

7.7.2 STATE OF THE ART OF BLOCKCHAIN

In open systems (the system where anyone has the authority to share as a peer without any restraint, and each peer is either untrusted or significantly strange), blockchain conceptualizes a state of the art that resolves the problems of DoS assaults, but it requires a significant expense for every exchange.

Individual algorithm with blockchain concept is a conglomeration of points of interest and weaknesses. Based on the nature of industry and the digital ecosystem constructed by user, the users can select the most appropriate protocol for their commercial infrastructure.

7.8 CLOUD STORAGE

Google Drive, Dropbox, Gmail, etc. are the emerging applications of cloud storage. Cloud storage permits end users to stockpile their data remotely and hence obtain it anywhere, anytime. Nowadays, we hoard all data in the digital format; it reduces effort of user in real time as conventionally files are maintained physically. Government sectors are also taking the advantages of cloud data store and utilizing the benefits of public and private clouds to store, retrieve, and manipulate transactions of data. Encryption of data is followed to store data over cloud, and cloud services assist all organizations under a scheme of Service-Level Agreements. It covers data integrity and privacy. Many companies are shifting their entire data center onto the cloud and enjoy the benefits of elasticity, load balancing, redundancy, availability, and integrity. Recently, this market is increasing at an exponential rate, with the majority of cloud storage being served by large companies such as Google, Microsoft, and Amazon. You may have observed that even mobile companies like Samsung and Apple now have their own cloud storage facilities, e.g., Samsung Cloud and iCloud, for end users [4, 23].

7.8.1 BLOCKCHAIN IN CLOUD STORAGE

Blockchain cloud storage remedial solutions allows the user's data and fragment it up into small chunks. After this, we add an additional layer of security and redistribute the fragmented packet throughout the network. The core idea behind blockchain features includes the hashing function, public/private key encryption, and transaction ledgers. Each fragment of datapacket is accumulated in a decentralized spatial location. It is a challenging task for intruders if they try to hack into it; the hacker first gets the encrypted data and then receives only a chunk of the data and not the entire file – this extends the mode of security for documents in blockchain–based cloud storage.

The second benefit of blockchain–cloud storage is that the owner is hidden as the node does not include the owner's information. The miner only receives a packet of data; hence, all the sensitive information is protected and secured. It extends the concept of data redundancy and load-balancing mechanisms, which are applied for high availability and quick access. At the same time, if a user tries to fetch data, all packets of the data are first validated, and if any sort of deformation is found on a data packet, then the miner who is responsible for altering the data packet is discarded from the network, and that altered part is matched from another redundant replica. Thus, all users get original and identical copies of the data.

Blockchain–cloud infrastructure epitomizes faster technological innovation. Blockchain–cloud infrastructure offers cryptographic blocks in blockchain. To realize this technology, we as researchers must ensure position of cloud in blockchain.

Data exchange over cloud needs an extensive secure routing protocol to reach central server. Due to massive proliferation of data, we need a blockchain logistic platform. It would support secure data processing over the Internet. Every cloud application goes through an audit when billions of cloud transactions enter into blockchain. It will scale the platform in real time in terms of security and scalability. Cloud–blockchain application is a multifaceted architecture to exchange scientific, technical, and academic data. It can ensure intellectual of property rights in current scenario. It can decentralize cloud infrastructure remotely. To provide more secure application, current generation needs blockchain and cloud together. This technology is gaining attention due to its encrypted ledger. When any enterprise adheres to blockchain–cloud technology, it provides more accountability. It is broker free, i.e., no intermediary control exists. Multiple-party verification in blockchain–cloud infrastructure facilitates storing sensitive information.

Integration of blockchain with cloud environments provides protection of data and security of information over network. Cloud computing provides high-end solution for technology virtualization. It can promote cost saving for an infrastructure building. Cloud computing offers platform independency, distributed computing environment, and on-demand resource access at a large scale. There are residing various challenging dimension in cloud computing, i.e., stored data over cloud are not secure. So, any unauthorized access can collapse the entire business ecosystem of an enterprise. As per resource accessibility over cloud, it can be categorized into three types: (i) public cloud, (ii) private cloud, and (iii) hybrid cloud. Public cloud ensures high scalability, reliability, and low-cost infrastructure building. Private cloud offers service within an organization. It is managed by an internal authority. It offers a distinct level of security for a high-end critical application. Hybrid cloud is a combination of public and private clouds. It is flexible as well as secure. In this chapter, we attempt to provide an overview of blockchain–cloud concept for a decentralized cryptographic model. We emphasized security application of blockchain and its significance in area of cloud computing. In the present scenario, blockchain can be integrated with any distributed generic cloud model to ensure quality of service and end-user security. Billions of transactions are being generated over cloud. It is a heterogeneous, sensitive, and unresponsive information. In order to deal with the above issue, a cloud–blockchain architecture is developed. Origin of this architecture is widely accepted for validating any transaction. Each node in blockchain receives a transaction id given by end user. This transaction index initiates the operation and proceeds until the process ends. This set of transaction is recorded in a transaction log or ledger. In blockchain, we maintain a queue data structure for all committed transactions. To update a transaction, mining node confirms an operation over blockchain network.

7.8.2 The Risks of Cloud Computing Are Increasing

Subsequently, the risks related to cloud computing are increasing in terms of data control and management. So, we need a blockchain–cloud model that will completely transform business transaction over cloud, as shown in Figure 7.2. The purpose of this model is to significantly identify potential risks behind cloud computing, and at

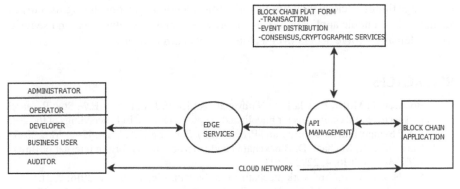

FIGURE 7.2   A model on cloud edge service and blockchain.

the same time, it can be overcome by blockchain technology. The model analyses the theoretical framework of blockchain–cloud application. It operates in a block-wise sequence. Here public network interacts with a cloud edge service, which is a cross cloud support framework that monitors the path between user and cloud resource. It ensures a better message control for a consistent high network performance. Cloud API (application programming interface) provides RPC (remote procedure call) interface. It offers cross cloud compatibility. It can integrate application and workload into cloud using these APIs. Some well-known APIs are SaaS API, PaaS API, and Iaas API.

CONCLUSION AND FUTURE SCOPE

Blockchain is recently analyzing the latest generic solution of every enterprise. The technology changes the entire gamut of transaction over network when integrated into another robust technology, cloud computing. It projects a scalable security to its end user. Cloud computing is a conglomeration of large networks and virtual data stores. When we amalgamate blockchain with cloud, it gives a new insight into storage, replication, and access to transactional database. It blends the security concept between task user and data collaborator over cloud. The future scope of this study is the research gap that exists between data modeling and orientation of blockchain–cloud. Converging all applications in blockchain incurs a high operational cost. Rationality phase of cloud–blockchain will obtain an optimal solution if distribution of cloud data can be analyzed in terms of bilinear pairing between cloud data store and blockchain ledger. We must focus on privacy and availability of data on blockchain–cloud creation in edge computing, where we put resource near to device location. It can effectively utilize all resources over cloud under constrained security. To achieve safe, secure, and scalable data sharing over blockchain–cloud model, we still wait for a relevant algorithm that will resolve any dishonest behavior prior to data movement over cloud. To improve latency period over cloud–blockchain model, we need a multiagent ledger learning algorithm that can control the infrastructure by prior learning. Blockchain–cloud integration provides a state of the art to manage things in cloud computing with utmost autonomy. Considering the current scenario, it is a noble application that can provide scalability, elasticity, and security for on-demand remote access over a cloud–blockchain network.

REFERENCES

1. Bonneau, J.; Miller, A.; Clark, J.; Narayanan, A.; Kroll, J.A.; Felten, E.W. Sok: Research perspectives and challenges for bitcoin and cryptocurrencies. In Proceedings of the 2015 IEEE Symposium on Security and Privacy (SP), San Jose, CA, USA, 17–21 May 2015.
2. Christidis, K.; Michael, D. Blockchains and smart contracts for the internet of things. *IEEE Access* 2016, 4, 2292–2303.
3. Shi, N. A new proof-of-work mechanism for bitcoin. *Financ. Innov.* 2016, 2, 31.
4. Paul, G.; Sarkar, P.; Mukherjee, S. Towards a more democratic mining in bitcoins. In Proceedings of the International Conference on Information Systems Security, Hyderabad, India, 16–20 December 2014; Springer International Publishing: Cham, Switzerland, 2014.
5. Il-Kwon, L.; Young-Hyuk, K.; Jae-Gwang, L.; Jae-Pil, L. The analysis and counter-measures on security breach of bitcoin. In Proceedings of the International Conference on Computational Science and Its Applications, Guimarães, Portugal, 30 June–3 July 2014; Springer International Publishing: Cham, Switzerland, 2014.
6. Beikverdi, A.; JooSeok, S. Trend of centralization in Bitcoin's distributed network. In Proceedings of the 2015 16th IEEE/ACIS International Conference on Software Engineering, Artificial Intelligence, Networking and Parallel/Distributed Computing (SNPD), Takamatsu, Japan, 1–3 June 2015.
7. Huang, H.; Chen, X.; Wu, Q.; Huang, X.; Shen, J. Bitcoin-based fair payments for out-sourcing computation of fog devices. *Future Gener. Comput. Syst.* 2016, 78, 850–858.
8. Huh, S.; Sangrae, C.; Soohyung, K. Managing IoT devices using blockchain platform. In Proceedings of the 2017 19th International Conference on Advanced Communication Technology (ICACT), Bongpyeong, Korea, 19–22 February 2017.
9. Armknecht, F.; Karame, G.; Mandal, A.; Youssef, F.; Zenner, E. Ripple: Overview and outlook. In *Trust and Trustworthy Computing*; Conti, M., Schunter, M., Askoxylakis, I., Eds.; Springer International Publishing: Cham, Switzerland, 2015; pp. 163–180.
10. Vasek, M.; Moore, T. There's no free lunch, even using bitcoin: Tracking the popularity and profits of virtual currency scams. In Proceedings of the International Conference on Financial Cryptography and Data Security, San Juan, Puerto Rico, 26–30 January 2015; Springer: Berlin/Heidelberg, Germany, 2015.
11. Zhang, J.; Nian, X.; Xin, H. A Secure system for pervasive social network-based health-care. *IEEE Access* 2016, 4, 9239–9250.
12. Singh, S.; Jeong, Y.-S.; Park, J.H. A survey on cloud computing security: Issues, threats, and solutions. *J. Netw. Comput. Appl.* 2016, 75, 200–222.
13. Kaskaloglu, K. Near zero bitcoin transaction fees cannot last forever. In Proceedings of the International Conference on Digital Security and Forensics (DigitalSec2014), The Society of Digital Information and Wireless Communication, Ostrava, Czech Republic, 24–26 June 2014.
14. Aitzhan, N.Z.; Davor, S. Security and privacy in decentralized energy trading through multi-signatures, blockchain and anonymous messaging streams. *IEEE Trans. Dependable Secur. Comput.* 2016, 3, 99.
15. Heilman, E.; Foteini, B.; Sharon, G. Blindly signed contracts: Anonymous on-blockchain and off-blockchain bitcoin transactions. In Proceedings of the International Conference on Financial Cryptography and Data Security, Christ Church, Barbados, 22–26 February 2016; Springer: Berlin/Heidelberg, Germany, 2016.

16. Natoli, C.; Gramoli, V. The blockchain anomaly. In Proceedings of the 2016 IEEE 15th International Symposium on Network Computing and Applications (NCA), Cambridge, MA, USA, 31 October–2 November 2016.

17. Swan, M. *Blockchain: Blueprint for a New Economy*; O'Reilly Media, Inc.: Sebastopol, CA, 2015.

18. Tschorsch, F.; Scheuermann, B. Bitcoin and beyond: A technical survey on decentralized digital currencies. *IEEE Commun. Surv. Tutor.* 2015, 18, 2084–2123.

19. Wressnegger, C.; Freeman, K.; Yamaguchi, F.; Rieck, K. Automatically inferring malware signatures for anti-virus assisted attacks. In Proceedings of the 2017 ACM on Asia Conference on Computer and Communications Security, Abu Dhabi, UAE, 2–6 April 2017.

20. Decker, C.; Roger, W. Information propagation in the bitcoin network. In Proceedings of the 2013 IEEE Thirteenth International Conference on Peer-to-Peer Computing (P2P), Trento, Italy, 9–11 September 2013.

21. Nakamoto, S. Bitcoin: A peer-to-peer electronic cash system. Available online: https://bitcoin.org/en/bitcoin-paper (accessed on 29 June 2017).

22. Bozic, N.; Guy, P.; Stefano, S. A tutorial on blockchain and applications to secure network control-planes. In 2016 3rd Smart Cloud Networks & Systems (SCNS), Abu Dhabi, UAE, 2016.

23. Bradbury, D. The problem with Bitcoin. *Comput. Fraud Secur.* 2013, 11, 5–8.

8 Solidity Essentials

Parv Garg and Neeraj Khadse
ICFAI Foundation for Higher Education
(Deemed to be University)

CONTENTS

8.1 INTRODUCTION

Solidity [1] is a statically typed, contract-oriented, high-level language that is used
to code smart contracts [2, 3], which are functions that control the transfer of funds
based on certain conditions. Solidity develops a machine-level code that executes
on Ethereum Virtual Machine (EVM) [4], a distributed blockchain-based platform.
It has many similarities with advanced programming languages like C, Python,
and JavaScript. It supports various libraries, complex user-defined data types, and
Object-Oriented Programming (OOP) concepts like inheritance, among several other
features. Some use cases of Solidity-coded smart contracts are systematic voting, auc-
tions, lottery, etc.

   In this section, we will focus on the basics of Solidity programming, and in the
upcoming sections, we will learn more about smart contracts and EVM.

8.2 ENVIRONMENT SETUP

There are quite a few platforms and methods to work with solidity language. Some of
them are:

   a. **Remix Online Editor**
      Online Editors, such as Remix, are recommended for small and quick proj-
      ects. It renders a good range of solidity versions. It provides an environment to
      code, compile, and deploy smart contracts. Remix online editor is available at:
         "https://remix.ethereum.org"
   b. **npm / Node.js**
      Solidity compiler, solcjs, is also installed by npm using the command:

      ```
      $sudonpm install -g solc
      ```

   c. **Docker Image**
      Docker Image can be fetched for solidity programming using the command:

      ```
      $docker pull ethereum/solc:stable
      ```

d. Binary Package
   Binary Packages are used to install the full-fledged solidity compiler.

8.3   GETTING STARTED

In this section, we will learn some basic elements of solidity language [5].
   Let us first take a look at Figure 8.1 and analyze it.

a. Pragma
   The first line of the code carries a pragma directive. It is practiced to specify the version of solidity. Here the version specified is "0.4.17." In a project, different files can have different versions; thus, a pragma directive is enacted in local to a source file and automatically ignored while importing.

b. Comments
   Comments are additional notes or annotations provided by the programmer. These are ignored by the compiler during interpretation and compilation. In solidity, comments are used just as in C++: // for single-line comment, and /* and */ for multi-line comments. As observed in Figure 8.1, comments are written after each statement to explain its purpose.

c. Keywords
   Before moving forward, let us take a glance at a few important Keywords in solidity shown in Table 8.1.
      Most of these work similar to any other advanced programming language.

```
1   pragma solidity ^ 0.4.17; //To Specify the version of solidity
2
3 ▾ contract SimpleStorage { //"SimpleStorage" is the name of the contract
4       uint storedData; //unsigned integer "storedData"
5 ▾     function set(uint x) public { //Public function with "set", take unsigned integer as an input
6           storedData = x;
7       }
8 ▾     function get() public view returns (uint) { //Public function with "get", view function does not make any changes in contract
9           return storedData;
10      }
11  }
```

FIGURE 8.1   Getting started code.

TABLE 8.1
Keywords in Solidity

Abstract	after	alias	apply
Auto	case	catch	copyof
Default	define	final	immutable
Implements	in	inline	let
Macro	match	mutable	null
Of	override	partial	promise
Reference	relocatable	sealed	sizeof
Static	supports	switch	try
Typedef	typeof	unchecked	

d. Import

Solidity like other languages supports import statements. Importing a file here is similar to that of JavaScript. However, the default export is not supported.

```
Import "filename";
```

This command will import all the global variables from the file, filename.

8.4 VALUE AND OPERATOR TYPES

8.4.1 Value Types

Solidity, like any other language, supports many built-in data types. Many value types in solidity are similar to those of C++. Let us take a glance at some listed in Table 8.2.

8.4.2 Variable Types

8.4.2.1 State Variables

Variables that are declared inside contract scope and outside of function scope are called "State variables." The values of state variables are stored permanently in contract storage. An example is shown in Figure 8.2.

TABLE 8.2
Value Types

Name	Description	Examples
string	Sequence of characters	"Hi there!", "Chocolate"
bool	Boolean value	True, False
int	Integer, positive or negativeHas no decimal	0, –3000, 59248
uint	"Unsigned" integer, positive number. Has no decimal	0, 3000, 9857
fixed/ufixed	"Fixed" point number. Number with a decimal after it	20.001, –42.4242, 3.14
address	Has methods tied to it for sending money	Ox18bae199c8dbae384d

```
1   pragma solidity ^0.4.17;
2 ▾ contract SolTest {
3       uint Data;       // State variable
4 ▾     constructor() public {
5           Data = 15;   // Using State variable
6       }
7   }
```

FIGURE 8.2   State variables.

8.4.2.2 Local Variables

Variables that are declared inside function scope are called "Local variables." These cannot be accessed outside the function. The values of local variables are not saved between different function calls. Notice the code shown in Figure 8.3 to understand local variables in solidity.

8.4.2.3 Global Variables

Variables that exist permanently in a global namespace are called "global or special variables." They provide details related to blockchain [6, 7] and other utility functions, i.e., transaction properties. Some block and transaction properties are shown in Table 8.3.

```
1   pragma solidity ^0.4.17;
2   contract SolTest {
3       uint Data;      // State variable
4       constructor() public {
5           Data = 15;   // Using State variable
6       }
7       function getAnswer() public view returns(uint){
8           uint x = 5;  // local variable
9           uint y = 6;
10          uint answer = a * b;
11          return answer; //access the local variable
12      }
13  }
```

FIGURE 8.3   Local variables.

TABLE 8.3
Block and Transaction Properties

Name	Returns
blockhash (uint blockNumber) returns (bytes32)	Hash of the given block – only works for 256 most recent, excluding current, blocks
block.coinbase (address payable)	Current block miner's address
block.difficulty (uint)	Current block difficulty
block.gaslimit (uint)	Current block gaslimit
block.number (uint)	Current block number
block.timestamp (uint)	Current block timestamp as seconds since unix epoch
gasleft () returns (uint256)	Remaining gas
msg.data (bytes calldata)	Complete calldata
msg.sender (address payable)	Sender of the message (current caller)
msg.sig (bytes4)	First four bytes of the calldata (function identifier)
msg.value (uint)	Number of wei sent with the message
now (uint)	Current block timestamp
tx.gasprice (uint)	Gas price of the transaction
tx.origin (address payable)	Sender of the transaction

8.4.2.4 Rules for Naming Solidity Variables

a. Variable must not be a keyword.
b. Variable must not start with a number (0–9); rather, it should always start with either an alphabet or underscore. For example, _abc, toCook are valid, but 2Cook is invalid.
c. Solidity is a case-sensitive language, i.e., "name" and "Name" are recognized as different variables.

8.4.3 OPERATOR TYPES

Operators are special symbols used for an arithmetic or logical operation. Operators in solidity are similar to that present in C++. In this section, we will review different operators and their uses.

8.4.3.1 Arithmetic Operators

TABLE 8.4
Arithmetic Operators

Operations	Symbol	Explanation	Example (X = 10, Y = 5)
Addition	+	Adds two numbers	X + Y, will return 15
Subtraction	–	Subtracts two numbers	X – Y, will return 5
Multiplication	*	Multiplies two numbers	X * Y, will return 50
Division	/	Divides two numbers	X / Y, will return 2
Modulus	%	Find modulus of two numbers	X % Y, will return 0
Increment	++	Increment two numbers	X ++, will return 11
Decrement	—	Decrement two numbers	X —, will return 9

8.4.3.2 Comparison Operators

TABLE 8.5
Comparison Operators

Operations	Symbol	Explanation	Example (X = 10, Y = 5)
Equal		Returns 1 if two operators are equal, else 0	X = = Y, will return false
Not equal	!=	Returns 1 if two operators are unequal, else 0	X != Y, will return true
Greater than	>	Returns 1 if the first variable is greater than second, else 0	X > Y, will return true
Less than	<	Returns 1 if the first variable is less than second, else 0	X < Y, will return false
Greater than equal to	>=	Returns 1 if the first variable is greater than or equal second, else 0	X >= Y, will return true
Less than equal to	<=	Returns 1 if the first variable is less than or equal second, else 0	X <= Y, will return false

8.4.3.3   Logical Operators

TABLE 8.6
Logical Operators

Operations	Symbol	Explanation	Example (X = 10, Y = 5)
Logical AND	&&	Returns true only when both operands are true, otherwise false	X && X, will return true X && Y, will return false
Logical OR	||	Returns true even when one operand is true, otherwise false	X && Y, will return true Y && Y, will return false
Logical NOT	!	Returns true even when one operand is true, otherwise false	!(X), will return false !(Y), will return true

8.4.3.4   Assignment Operators

TABLE 8.7
Assignment Operators

Operations	Symbol	Explanation	Example (A = 10 and B ≡ 5)
Simple assignment	=	Assigns a value to a variable	A = B, now the value of both A and B is 5
Add and assignment	+=	Adds and assigns to a variable	A += B, means A = A + B, which is 15
Subtraction and assignment	−+	Subtracts and assigns to a variable	A −= B, means A = A − B, which is 5
Multiplication and assignment	*=	Multiplies and assigns to a variable	A *= B, means A = A * B, which is 50
Division and assignment	/+	Divides and assigns to a variable	A /= B, means A = A / B, which is 2
Modulus and assignment	%=	Finds Modulus and assigns to a variable	A %= B, means A = A%B, which is 0

8.4.3.5   Conditional Operator (?: (conditional))

If condition is true? The value X : Otherwise Y.

For Example:

i. A = (5==5)? 10 : 20, here the value of A will be assigned 10, as the condition is true.

ii. B = (5==10)? 10 :20, here the value of B will be assigned 20, as the condition is false.

8.5   CONTROL STRUCTURES

Solidity establishes control structures like other curly bracket languages. It implements if, else, else if, while, do, for, continue, and break semantics similar to C or JavaScript.

Let us understand the syntax for each type.

8.5.1 Loop

Loops are used where the repetition of some codes is required; it is programmed using for, while, or do while. For example,

```
while (condition)
{
/*
These statements within these curly braces will execute
repeatedly until the condition fails
    */
}
```

Likewise, for and do while loops are used.

8.5.2 Decision-Making

Decision-making tools are required to determine distinct paths based on particular conditions. In Solidity, if, else, or else if are used for decision-making. For example,

```
if (condition_1)
{
//Execute statements of this block
//Ignore all other else if or else belonging to this if
}
else if (condition_2)
{
//Execute statements of this block
//Ignore all other else if or else belonging to this if
}
...
//Multiple else if are possible
else //Works only if all the above if or else if condition
fails
{
//Execute statements of this block
}
```


8.6 DATA STRUCTURES

8.6.1 Array

Often, we have discussed arrays as a data type, but categorically it is a data type that stores a collection of data [8]. Hence, an array is a collection of variables applying to the same data type. In an array, a search operation is usually linear time associated, as for each additional record the search time increases slightly. Generally, array in solidity is of two types: fixed array and dynamic array.

8.6.1.1 Fixed Array

In solidity, the most primitive array type is a fixed array. It is an array with a fixed length. In JavaScript, arrays can be resized by appending or removing a record, but the length of the fixed array, in solidity, cannot be altered. Its size is predefined at declaration.

```
Syntax: type arrName [ arrSize ]
Example: uint storage [10]
```

8.6.1.2 Dynamic Array

A dynamic array can change its size over time. Dynamic array functions similar to JavaScript array, i.e., we can modify its length by adding or removing an element. Its size is not predefined at declaration.

```
Syntax: type[ ] arrName
Example: uint[ ] storage
```

While declaring a dynamic array, we do not specify its length in the square brackets. This implies that we are designing an array whose length can stretch or shrink according to the requirement [6].

8.6.1.3 Array Properties

Arrays in solidity hold three main properties: length, push, and index.

a. **Length**

The length property returns the size of an array. It can only be modified by dynamic arrays.

```
Syntax: arrName.length;
Example: balance.length;
```

b. **Push**

The push property is used to append new elements at the end of an array. It can only be used for dynamic arrays.

```
Syntax: arrName.push();
Example: balance.push(100);
```

c. **Index**

The index property is practiced for reading particular elements in an array. It is supported by all types of arrays.

Refer to the example shown in Figure 8.4 to understand the functioning and characteristics of arrays.

8.6.2 MAPPING

A mapping is a collection of key-value pairs. Think of it as JavaScript objects, Ruby hashes, or Python dictionary. The key thing to keep in mind about mapping is that all

```
 1   pragma solidity ^0.4.17;
 2
 3▾  contract Test {
 4       uint[] public Array;
 5
 6▾      function Test() public {
 7           Array.push(5);
 8           Array.push(10);
 9           Array.push(30);
10       }
11
12▾      function getArrayLength() public view returns(uint) {
13           return Array.length;
14       }
15
16▾      function getFirstElement() public view returns(uint) {
17           return Array[0];
18       }
19   }
```

FIGURE 8.4   Array properties.

the keys and its values must be of the same data type. Mapping structure can operate on value types, as mentioned earlier as an integer, boolean, struct, and address. It holds two main parts: a keyType and its valueType [9].

```
Syntax:  mapping (keyType =>valueType) mapName
Example: mapping (int =>storedData) storage
```

A search inside of mapping is referred to as constant search time because no matter how many pieces of data we are storing inside the mapping, it is always going to take the same amount of time to execute a search.

8.6.2.1   Mapping Gotchas

A. In mappings, keys are not stored, which means that we cannot get a list of keys in mapping. Let's understand this using an example:

```
const info = {
name: 'John',
  age: 21,
  country: 'India'
};
Object.keys(info);
```

Output :

```
["name", "age", "country"]
```

This is a normal JavaScript object that lists out all information about a person. We can easily retrieve all the keys and values of an object. But the same is not true in the case of mapping in solidity. In solidity, we work with a classic data structure called "hash table." The hash table has a lookup process, as explained in Figure 8.5.

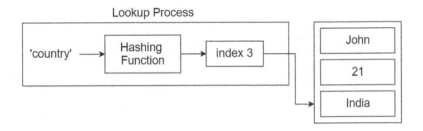

FIGURE 8.5 Lookup process (1).

Here on the right side, we can observe the underlying data structure that represents mapping. To look up a value from the mapping, we have to provide a key before time. After providing the key, it passes through a hashing function, and then the hashing function outputs some predefined index. So, when searched for "country" solidity point at index 3 inside mapping and retrieve the value of that element. Consequently, keys are not stored in mapping, i.e., we cannot prepare a list of keys within the mapping [7].

B. Also, values are non-iterable in a mapping. In other words, we cannot loop through mapping and print out the different values it contains. We cannot run a function or make a call to the mapping to retrieve all its values. This is closely related to the fact that we discussed before, the inability to access the list of keys in mapping. So, we can assert that mappings are only good for single-value look-ups and not for storing information which we need to iterate over.

C. In mapping, "all values exist." To understand this, let us take the same object "info," convert it to mapping, now try to lookup for a key, and say "state." It will then get passed to a hashing function and will return some index. Rather than informing us of no value at that particular index, it will return a default value for that element. For example, if all the values in the "info" object were "string," then the default value would be an empty string. The default value that is returned from the mapping depends on the value type passed to the mapping. This concept is illustrated in Figure 8.6.

Refer to the code in Figure 8.7 to understand the working of mappings in solidity.

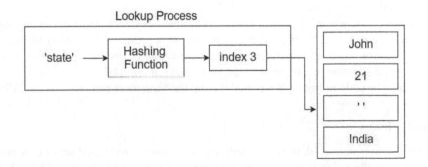

FIGURE 8.6 Lookup process (2).

```
1    pragma solidity ^0.5.0;
2
3 ▾  contract AccountBalance {
4        mapping(address => uint) public balance;
5
6 ▾      function updateBal(uint newBal) public {
7            balance[msg.sender] = newBal;
8        }
9    }
10 ▾ contract Update {
11 ▾     function updateBal() public returns (uint) {
12            AccountBalance accountBalance = new AccountBalance();
13            accountBalance.updateBal(15);
14            return accountBalance.balance(address(this));
15        }
16   }
```

FIGURE 8.7 Mappings in Solidity.

8.6.3 STRUCT

A struct is a collection of key-value pairs that can hold different value types. It is generally used to represent a singular entity.

Syntax for struct:

```
struct structName {
        type1 typeName_1;
        type2 typeName_2;
        type3 typeName_3;
        }
```

Let us understand with an example.

```
struct Car {
     string make;
     string model;
uint value;
    }
```

Here in this example, we have designed a struct, car, which contains three attributes make, model, and value.

At this point, you might notice some resemblance between struct and mappings, although there is a difference. Mapping is used to represent a collection of things, whereas a struct is used to represent a singular thing. For instance, in mapping, we can store a collection of cars rather than just a singular car [1].

Notice the code in Figure 8.8 to understand the functioning of structs in solidity.

8.6.4 ENUM

Enum is a user-defined data type in solidity. It contains a restricted set of constants from which the value is assigned to a variable. Enum limits the variable to hold any unspecified value, thus reducing bugs. Suppose you are creating a clothing

```
1   pragma solidity ^0.5.0;
2
3 ▾ contract Test {
4 ▾     struct Car {
5           string make;
6           string model;
7           uint value;
8       }
9       Car car;
10
11 ▾    function setCar() public {
12          car = Car('Audi', 'A6', 5000000);
13      }
14 ▾    function getCarValue() public view returns (uint) {
15          return car.value; // Accessing value from struct Car using (.) operator
16      }
17  }
```

FIGURE 8.8   Structs in Solidity.

application, then you would prefer to restrict the size selection to small, medium, and large. Therefore, eliminating the chances of erroneous input.

The code in Figure 8.9 illustrates the concept of enum in Solidity.

8.6.4.1   Enum Gotchas

a. Numbers are not accepted as a constant in an enum. And for booleans, the first letter must be capital (True or False).

b. Enum identifies its End of Declaration by the closing curly bracket. A semicolon (;) is used to indicate the end of single-line instructions, and curly brackets ({}) are used to indicate the start or end of a set of instructions like functions, struct, modifiers, enums, and the contract itself. Therefore, we need not specify the End of Declaration with a semicolon.

8.6.5   STRINGS

The string data type is similar to an array of characters. Solidity sustains both single (') and double quote (") string literals. The "string" keyword can be used to declare a string variable type. Although it is recommended using byte-type literals compared to

```
1   pragma solidity ^0.5.0;
2
3 ▾ contract Direcion {
4       enum DirectionChoices { GoLeft, GoRight, GoStraight, StandStill };
5       function Direction()
6 ▾     {
7           choices = DirectionChoices.GoStraight;
8       }
9       function getChoice() returns (uint ch)
10 ▾    {
11          ch = uint256(choices);
12      }
13      DirectionChoices choices;
14  }
```

FIGURE 8.9   Enum in Solidity.

```
1  pragma solidity ^0.5.0;
2
3 ▾ contract Notes {
4     string note = "Hello!";
5  }
```

FIGURE 8.10 Strings in Solidity.

```
1  pragma solidity ^0.5.0;
2
3 ▾ contract Notes {
4     bytes32 note = "Hello!";
5  }
```

FIGURE 8.11 Bytes in Solidity.

string-type literals, as string operations require higher gas [8, 9]. Strings are implicitly convertible to byte types. We apply string for data that is longer than 32 bytes.

In the example in Figure 8.10, note is a string variable and "Hello!" is a string literal. However, as discussed, we should use the byte type as string type is an expensive operation.

In the example in Figure 8.11, we have used byte32 instead of string type.

8.7 FUNCTION

A function is a block of code that can be used multiple times. It lessens the necessity to rewrite the same code afresh. It helps to divide a big program into small sections, and hence facilitates modularity. Functions in solidity are implemented alike other advanced programming languages.

8.7.1 Function Definition

Initially, a function is defined after which it can be called "multiple times." Usually, the function keyword is adopted to define a function followed by a distinct function name, a list of parameters within parentheses (could be empty), and a statement block inside curly brackets.

```
Syntax: function functionName (parameterList) scope returns () {
        // statements
        }
```

In the example given in Figure 8.12, a function, getAnswer(), is defined without any parameters.

```
1    pragma solidity ^0.5.0;
2
3 ▾  contract Multiply {
4 ▾      function getAnswer() public view returns(uint){
5            uint x = 5; // local variable
6            uint y = 10;
7            uint answer = x * y;
8            return answer;
9        }
10  }
```

FIGURE 8.12   Function definition.

8.7.2   FUNCTION CALL

To call or invoke a function inside a contract, we specify the name of that function followed by required parameters inside the parenthesis.

Notice the code in Figure 8.13 to understand function calls in solidity.

8.7.3   FUNCTION PARAMETERS

In the previous example, function parameters were not needed. However, a function may accept multiple parameters. Function parameters are values passed to a function during function calls. With the help of these parameters, a function processes code. A function may need several parameters which can pass inside parentheses separated by commas.

In the example given in Figure 8.14, x and y are function parameters.

8.7.4   RETURN STATEMENT

To return a variable from a function, the "return" keyword followed by the variable name is written. Usually, the return statement is written at the end of a function definition.

```
1    pragma solidity ^0.5.0;
2
3 ▾  contract Multiply {
4 ▾      function getAnswer() public view returns(uint) {
5            uint x = 5;
6            uint y = 10;
7            uint answer = x * y;
8            return getMarks(answer);
9        }
10 ▾      function getMarks() public {
11           // some code
12       }
13  }
```

FIGURE 8.13   Function call.

```
1    pragma solidity ^0.5.0;
2
3 ▾  contract Multiply {
4        uint mul;
5 ▾      function getAnswer(uint x, uint y) public {
6            mul = x * y;
7        }
8    }
```

FIGURE 8.14   Function parameters.

In the previous example, we have assigned variable mul as a product of x and y. Now, what if we want to return its value? We can do so by writing a return statement at the end of the function. Refer to the example shown in Figure 8.15.

8.7.5   FUNCTION MODIFIERS

The function modifier enables us to control the behavior of a function. It is used for a condition check before the function execution. It reduces the amount of duplicate code or logic. The special symbol "_;" is mandatory for a modifier definition. At "_;" the function source is merged with the modifier code.

Let's understand with the example in Figure 8.16.

```
1    pragma solidity ^0.5.0;
2
3 ▾  contract Multiply {
4        uint mul;
5 ▾      function getAnswer(uint x, uint y) public {
6            mul = x * y;
7            return mul;
8        }
9    }
```

FIGURE 8.15   Return statement.

```
1 ▾  contract Owner {
2        address ownerAddress;
3 ▾      constructor() {
4            ownerAddress = msg.sender;
5        }
6        // Can only be called by the owner
7 ▾      function test() public onlyOwner {
8
9        }
10       // Check if the function is called by the owner of the contract
11 ▾     modifier onlyOwner() {
12           require(msg.sender == ownerAddress);
13           _;
14       }
15   }
```

FIGURE 8.16   Function modifiers.

8.7.6 FUNCTION TYPES

In solidity, several function types affect the behavior and execution of a function. Let's discuss them one by one.

A. **Public**: Public functions can be accessed directly from outside. They develop as members of the smart contract interface. They can be called from within or even externally.

B. **Private**: Private functions can only be called inside the contract. They are inaccessible by the other contracts, not even within the derived contracts. Unlike the public, these functions are not members of the smart contract interface.

C. **Internal**: Internal functions like private cannot be accessed from outside. They can only be used within the contract they are defined in or with the contracts inheriting from it, without using "this" keyword. These functions are also not members of the smart contract interface.

D. **External**: External functions can be accessed directly externally, but not internally. They can be accessed easily by other contracts or via transactions [10]. These functions are members of the smart contract interface.

E. **View**: View functions are those that remain unchanged. So, we can use the terms "view" and "constant" interchangeably. View function returns data but does not modify the contract data.

F. **Pure**: Pure functions cannot modify or read the contract's data. They are unauthorized to read state or transaction variables [11].

G. **Payable**: Payable functions are utilized to accept payment in the form of ethers. Whenever a payable function is used by some outside entity, it attempts to send some money with it. This function call will fail if required ethers [11] are not sent.

8.7.7 FALLBACK FUNCTION

A fallback function is a nameless function. It neither accepts an argument nor returns any data. It is evoked when a contract receives anonymous ethers or when the given function identifier is undefined. A smart contract can maintain only one fallback function. If a contract is expected to receive some ethers, then its fallback function must include the "payable" keyword.

Let's understand this with the example in Figure 8.17.

In the above example, we have defined a contract "Test." It holds a fallback function that increments x value for each invalid function call. The contract "Caller" shows how to call a function that does not exist, incrementing the value of x on each call.

8.7.8 FUNCTION OVERLOADING

Solidity supports function overloading. Function overloading is a feature that supports two or more functions with the same name but different functionalities. It occurs when functions with the same name have different return types or a number of arguments [5].

Refer to the example in Figure 8.18 to understand function overloading in solidity.

```
1    pragma solidity ^0.4.23;
2
3 ▼  contract Test {
4      uint x;
5 ▼    function() external payable {
6        x = x + 1;
7      }
8 ▼    function get() public view returns (uint) {
9        return x;
10     }
11   }
12
13 ▼  contract Caller {
14 ▼    function callTest(address testAddress) public {
15       testAddress.call("invalid");
16     }
17   }
```

FIGURE 8.17 Fallback function.

```
1    pragma solidity ^0.5.0;
2
3 ▼  contract Overload {
4 ▼    function getResult(uint x, uint y) public pure returns(uint){
5        return x + y;
6      }
7 ▼    function getResult(uint x, uint y, uint z) public pure returns(uint){
8        return x + y + z;
9      }
10 ▼   function callResultOne() public pure returns(uint){
11       return getSum(5,10);
12     }
13 ▼   function callResultTwo() public pure returns(uint){
14       return getSum(5,10,15);
15     }
16   }
```

FIGURE 8.18 Function overloading.

8.7.9 CRYPTOGRAPHIC FUNCTIONS

Cryptographic functions [12] are advanced functions employed to encrypt some data.
Solidity has built-in cryptographic functions; some of them are displayed in Table 8.8.
    The example in Figure 8.19 will help you further understand cryptographic functions.

8.8 CONTRACTS

In solidity, a contract [13] is a collection of functions and states. There can be multiple
contracts in a smart contract. Each contract resides at a distinct address on the Ethereum
blockchain [14–16]. Contract in solidity behaves alike class in OOP languages.
    Inheritance of contracts is also supported by Solidity. Like other classic OOP
languages, an inherited contract is called a "parent contract," and a contract that
inherits itself from a parent contract is called a "child contract." A parent contract is

TABLE 8.8
Cryptographic Functions

Functions	Return Type	Description
keccak256 (bytes memory)	bytes32	Takes the input, calculates the keccak256 hash.
sha256 (bytes memory)	bytes32	Takes the input, calculates the sha256 hash.
ripemdl60 (bytes memory)	bytes20	Takes the input, calculates the ripemdl60 hash.

```solidity
1   pragma solidity ^0.5.0;
2
3 ▾ contract Encrypt {
4 ▾     function encryption() public pure returns(bytes32 result){
5           return keccak256("XYZ");
6       }
7   }
```

FIGURE 8.19 Cryptographic function.

also known as a base contract. Inheritance is used to reduce the repetitive logic and duplicate code inside of a contract, thus improving the reusability of code.

For example, take the program in Figure 8.20.

A contract has the following properties.

8.8.1 Constructor

It is a special function that is declared with the "constructor" keyword. In a contract, a constructor is used to initialize the state variables. It gets invoked whenever a new contract is created. Only one constructor is allowed per contract.

```solidity
1   pragma solidity ^0.4.0;
2
3 ▾ contract Storage {
4       uint Data;
5
6 ▾     function setData(uint x) public {
7           Data = x;
8       }
9
10 ▾    function getData() public view returns (uint) {
11          return Data;
12      }
13  }
```

FIGURE 8.20 Contract in Solidity.

8.8.2 STATE VARIABLES

These are used to store the state of the variables. We have discussed state variables in the previous sections of this chapter.

8.8.3 FUNCTION

Function changes the state of the contract by modifying the state variables. We have also discussed functions in the previous sections of this chapter.

8.9 EVENTS

As we know, blockchain transactions are stored in blocks. Each transaction has some logs associated with it. A log records details of an event as they occur. An event in solidity holds the current status of the contract. It is stored in the form of logs. Events are declared globally in a contract and will be triggered from within its function. To declare an event, use the keyword "event" followed by an identifier and a list of parameters. And to emit an event, use the "emit" keyword followed by the event name and the arguments in parentheses.

Let's understand this with the help of the example in Figure 8.21.

8.10 ERROR HANDLING

Contracts in Solidity are compiled into bytecode. While writing contracts, we come across various errors. Unusual errors may occur at runtime or design time. The design level checks for any syntax errors while compiling. But it is harder to detect any runtime errors as they occur during the execution of a contract. Thus, it is crucial to check a contract for possible runtime errors. Hence, for reducing runtime and design time errors, a contract should follow a clean and robust fashion.

Solidity supports various methods for error handling. Some of them are listed in Table 8.9.

CONCLUSION

The solidity language is developed as a need for deploying smart contracts on the Ethereum Network. Solidity, a contract-oriented, high-level language, is used for writing smart contracts easily and efficiently. It develops a machine-level code that

```
1   pragma solidity ^0.5.0;
2
3   contract TestEvent {
4       event Deposit(address indexed _from, bytes32 indexed _id, uint _value);
5       function deposit(bytes32 _id) public payable {
6           emit Deposit(msg.sender, _id, msg.value);
7       }
8   }
```

FIGURE 8.21 Events in Solidity.

TABLE 8.9
Error Handling Methods

Methods	Description
assert (bool condition)	If the condition is not satisfied, an invalid operation code is caused by the method call and any changes to the state get reverted.
require (bool condition)	If the condition is not satisfied, this method call is reverted back to the original state.
require (bool condition, string memory message)	For errors in inputs or external components, this method is used. If the condition is not satisfied, this method is reverted back to the original state.
revert ()	This method reverts the changes done to the state and aborts the execution.
revert (string memory reason)	This method reverts the changes done to the state, aborts the execution, and provides an option for custom message.

executes on an EVM. Solidity has its similarities with other advanced programming languages like C, Python, and JavaScript. Most of its values and operator types are similar to C, with a few additional. Its control statements and functions are written like JavaScript. It supports data structures such as enum, array, struct, string, and mapping. OOP concepts like inheritance are present, which are useful for code reusability. Lastly, it also supports various libraries, complex user-defined data types, and several other features. In solidity, a contract plays a very crucial role. When the code written in solidity is deployed on a district address, it behaves as a smart contract. A contract is like a class in other OOP languages. With this, the chapter concludes.

REFERENCES

1. Dannen, C.: Solidity programming. In Chris Dannen (ed.) *Introducing ethereum and solidity.* Springer: Berlin, Germany, pp. 69–88, 2017.
2. Clack, C.D., Bakshi, V.A., Braine, L.: Smart contract templates: foundations, design landscape and research directions. *CoRR* abs/1608.00771. 2016.
3. Szabo, N.: Formalizing and securing relationships on public networks. *First Monday* 2(9), 1–28, 1997.
4. Dannen, C.: *Introducing ethereum and solidity: foundations of cryptocurrency and blockchain programming for beginners.* Apress, New York. p. 30.
5. Solidity official documentation. https://solidity.readthedocs.io/. 2017.
6. Tschorsch, F., Scheuermann, B.: Bitcoin and beyond: a technical survey on decentralized digital currencies. Cryptology ePrint Archive, Report 2015/464, 2015.
7. Bonneau, J., Miler, A., Clark, J., Narayanan, A., Kroll, J. A., and Felten, E. W. Research perspectives and challenges for Bitcoin and cryptocurrencies. Cryptology ePrint Archive, Report 2015/261, 2015.
8. Tikhomirov, S., Voskresenskaya, E., Ivanitskiy, I., Takhaviev, R., Marchenko, E., Alexandrov, Y.: SmartCheck: static analysis of ethereum smart contracts. In Sergei Tikhomirov, Ekaterina Voskresenskaya, Ivan Ivanitskiy, Ramil Takhaviev, Evgeny Marchenko, and Yaroslav Alexandrov (ed.) *1st international workshop on emerging trends in software engineering for blockchain.* ACM, Gothenburg, Sweden, pp. 9–16, 2018.
9. Wood, G.: Ethereum: a secure decentralised generalised transaction ledger. www.gavwood.com/paper.pdf. 2014.

10. Allison, I.: "Microsoft adds Ethereum language Solidity to Visual Studio." *International Business Times*. https://www.ibtimes.co.uk/microsoft-adds-ethereum-language-solidity-visual-studio-1552171. Last accessed September 1, 2020, March 30, 2016.

11. Waters, R.: 'Ether' brought to earth by theft of $50m in cryptocurrency. *Financial Times*, June 18, 2016.

12. Gatteschi, V., Lamberti, F., Demartini, C., Pranteda, C., Santamaría, V.: Blockchain and smart contracts for insurance: is the technology mature enough? *Future Internet* 10, 20 2018.

13. Tapscott, D., Tapscott, A.: *The blockchain revolution: how the technology behind bitcoin is changing money, business, and the world.* Penguin Publishing Group: London, pp. 72, 83, 101, 127. 2016.

14. Luu, L., Chu, D.H., Olickel, H., Saxena, P., Hobor, A.: Making smart contracts smarter. In: ACM Conference on Computer and Communications Security 2016, 254–269. 2016.

15. Buterin, V.: Ethereum: a next generation smart contract and decentralized application platform. https://github.com/ethereum/wiki/wiki/White-Paper, last accessed December 1, 2020. 2013.

16. Announcement of imminent hard fork for EIP150 gas cost changes. https://blog.ethereum.org/2016/10/13/announcement-imminent-hardfork-eip150-gas-cost-changes/ 2016.

9 Installing Frameworks, Deploying, and Testing Smart Contracts in Ethereum Platform

Tushar Sharma
ICFAI Foundation for Higher Education
(Deemed to be University)

CONTENTS

9.1 INTRODUCTION

Smart contract [1,2] is a piece of code with embedded business logic and rules of an agreement and stores assets/currency. The sole purpose of the smart contract is to move assets/currency, represent ownership that is stored in it based on some conditions. The developer of the smart contract imbibes a set of rules/logic to handle the state of the asset stored in it. Smart contracts are stored inside the blocks in the blockchain [3,4] network, due to which they inherit the properties of the blockchain technology, i.e., immutable [5] and distributed. So, once a smart contract is written, it is almost impossible to tamper it.

Smart contracts dwell in Ethereum [6,7] blockchain. Besides tracking the transactions made, smart contracts program the transactions too [8]. Smart contracts are written by Solidity programming language. Solidity [9] is a contract-oriented, high-level language for creating the smart contracts. It is a statically typed language. Solidity is compiled to bytecode that is executable on the Ethereum Virtual Machine (EVM) [7,10]. The execution of mathematical computations of smart contracts incurs "Gas-Cost".

9.2 GAS AND TRANSACTION

Gas is a system to quantify how much work we are executing with the code.

In order to get someone else to run our contracts we have to pay the money. This money is reflected in terms of gas. Whenever we issue a transaction that is meant to modify blockchain in any fashion or run any code, we have to specify two properties of our transaction object. The two properties are listed below:

1. gasPrice: Amount of Wei (small unit of ether) the sender is willing to pay per unit gas for handling a particular transaction.
2. startGas/gasLimit: Unit of gas that a particular transaction can consume.

The author of smart contract must specify the gas limit so that the validators can decide to mine it before validating the smart contract. If the reward of the smart contract is satisfactory, then miner chooses to invoke smart contract function call in their next block.

9.2.1 TRANSACTION CONSENSUS

A transaction is a consensus between any two parties willing to exchange assets, products, or services. Ethereum helps in accomplishing the transaction.

TABLE 9.1
Properties of a Transaction in Ethereum Platform

Nonce	Number of many times the sender has sent a transaction
To	Address of the account the money is transferring to
Value	Amount of the ether to be send to the desired address
gasPrice	Amount of ether the sender is willing to pay per unit gas for handling a particular transaction
startGas/gasLimit	Unit of gas that a transaction can consume

There are three types of transactions that can be executed in Ethereum platform:

1. Transfer of ether from one account to other accounts.
2. Deployment of a smart contract.
3. Invoking a function within a contract.

A transaction has some important properties related to it, as illustrated in Table 9.1.

The v, r, and s account properties relate to cryptographic [11] pieces of data that can be used to generate the sender's account address.

9.3   SMART CONTRACT ACCOUNT

Smart contracts can be considered as a way to automate some of the conditions and obligations described in the contract. Every smart contract has its own address and is useful while sending or receiving the ether. Smart contracts can identify the person who calls it, by this we can impose the privileged access to the contract.

Every smart contract account has four properties associated with it. These three properties are field, balance, storage, and code. Table 9.2 gives the description of each of these properties.

9.3.1   SMART CONTRACT STRUCTURE

The code shown in Figure 9.1 depicts the basic structure of smart contract.

The following are the features of above written smart contract:

1. The first line simply specifies that the source code is written for Solidity version 0.5.2. This is to make sure that the contract does not behave abnormally with a new compiler version. The keyword pragma can be used

TABLE 9.2
Properties of Smart Contract account

Field	Description
Balance	Number of ether any particular account possess
Storage	Data storage for the contract
Code	Raw machine code for the contract

```
pragma solidity ^0.5.2
contract Details{
    string public info;
    function Details(string initialInfo)  public {
    info = initialInfo;
    }
    function getInfo()public view returns (string){
    return info; } }
```

FIGURE 9.1   Details smart contract.

to enable certain compiler features or checks for a particular version of compiler.

2. Contract Details {}: This is a method to define a new contract using the keyword contract. Here, the name of the smart contract is Details. The new contract will have some number of methods and variables.

3. string public info: In this line, a variable "info" is defined of type "string," "public" is a keyword that indicates the access modifier or who can access the contents of this variable.

4. A storage variable will be automatically stored with the contract on the blockchain.

5. function declaration:

⇨ function getInfo() publicview returns (string)
{ return info;
}

The name of the function is getInfo() with "public view" as function type and "returns(string)" denotes the return type of the function.

9.4   DEPLOYMENT OF THE CONTRACT IN ETHEREUM RINKEBY TEST NETWORK

This section serves to introduce the reader to the principal motivation and also outlines the major concepts covered in this chapter.

Smart contract file is saved with the extension.sol file. Most of the editors such as Atom and VSCode have the support for the solidity language. Figure 9.2 represents that the smart contract when compiled gives two objects, namely bytecode and ABI [8,12]. On deploying a smart contract to Ethereum network, the bytecode of smart contract will be stored in the blockchain, and it returns the address of the smart contract called contract address. Just with the help of bytecode it's highly impossible to know the functionality of a smart contract, hence ABI plays a crucial role.

ABI is abbreviated as Application Binary Interface. ABI is our JavaScript interpretation layer needed to access the bytecode. It is required to interact with the

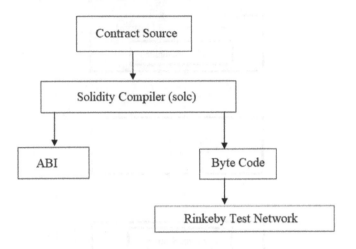

FIGURE 9.2 Compilation of sol file.

deployed smart contract using contracts address. If any account wants to invoke a function of smart contract, it utilizes ABI to hash the function and build EVM bytecode to invoke the function. It is a JavaScript formatted object (JSON) that depicts various functionalities of contract such as input/output parameters, name of functions, payability (whether ether can be sent or not), state mutability of functions. It facilitates the users to create appropriate messages for contract.

It's only the bytecode that is feed to Rinkeby test network for deploying and testing purpose of the smart contract.

Remix is a browser-based Solidity Integrated Development Environment (IDE) for coding, deploying, and testing the smart contracts. Various execution environments such as JavaScript VM, Injected web3 [13], and Web3 Provider are listed in Remix.

Remix is not just a code editor but it also holds a miniature false Ethereum network that can be used for deploying and interacting with the contact. Figure 9.3 illustrates how testing is done with Remix IDE.

After compilation of contract in Remix, click on "run" tab in Remix IDE in order to deploy the contract. As mentioned above, various execution environments such as JavaScript VM, Injected web3 [13], and Web3 Provider are listed in Remix IDE.

JavaScript VM– This environment allows the developer to run the contract directly using a JavaScript implementation of the EVM. In addition to this, the environment doesn't allow others to interact with the contract. So, this environment is good simple testing purposes.

Injected Web3 – The web3 library connects the Ethereum network from an application. By using the MetaMask, a chrome browser extension, we can interconnect with Ethereum network. This option will allow the developer to use the injected execution for deploying the contract to a test network or the main Ethereum network.

Web3 Provider– It connects directly to an Ethereum node via Hypertext Transfer Protocol (HTTP). So, this environment can be used to connect with Ganache module accounts.

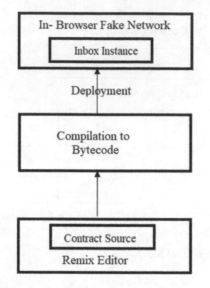

FIGURE 9.3 Testing with Remix.

For initial development and testing of contract, the JavaScript VM is the best option. But on the other hand, if we want others to interact with our contract, we'll need to use other available environment.

9.5 WEB3 LIBRARY

For creating and deploying the contract on to Ethereum network, Node.js and its modules are used. For compiling a solidity file in node.js a node module "solc" has to be installed on the machine. Using this module, the node.js will compile the solidity file, which gives the bytecode and the ABI. To interact with contract on the block-chain, web3 [13,14] library is used.

As mentioned Section 9.4, web3 library is used to connect to the Ethereum net-work from an application. So, web3 is used as the perfect solution for communicating between a JavaScript application and the Ethereum network. The ABI is given to the web3 library, which it uses to give a programmatic access to the contract deployed on the Ethereum network [13].

To utilize Web3, an object of Web3 class must be instantiated.

```
const Web3=require('web3');
   const web3=new Web3();//Instance
```

Multiple instances of web3 library can be created, and each instance is used to connect to a different Ethereum networks. To connect or communicate with an Ethereum network, the instance of web3 requires a communication layer known as

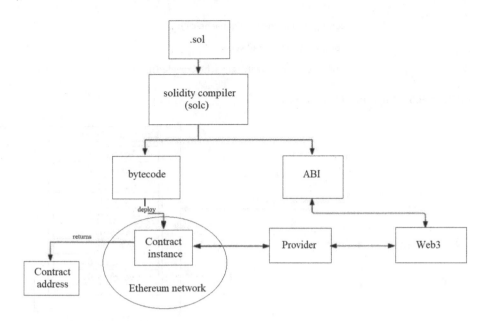

FIGURE 9.4 Deploying schemes.

provider illustrated in Figure 9.4. There are different types of providers and each can be used to reach out different Ethereum networks.

Every provider has a set of identical methods to send or receive a request from Ethereum network. These methods allow the web3 library to essentially send a request over to a local network and to receive a response to that request. The provider acts as a medium between the web3 library and the Ethereum network, just like when two persons are talking to each other, the air acts as a medium to transfer the sound waves from one person to another person.

Where the provider=ganache.provider();//Local Test Network

```
const web3= new Web3(provider);
```

9.5.1 GANACHE MODULE

The Ganache is a local Ethereum test network which can be installed through the "ganache-cli" npm module. It has to be required or imported into the project to create a local Ethereum test network. Figure 9.5 shows the code required to import Ganache module.

When the contract is deployed in a local network, there is no need for security and public and private keys to unlock the accounts that are provided by test network (Ganache). The good thing about these accounts is that they are created in an unlocked state and there is no requirement for ether. However, in case of an

```
const ganache = require('ganache-cli');

const Web3 = require('web3');

const web3 = new Web3(ganache.provider());
```

FIGURE 9.5   Importing Ganache module.

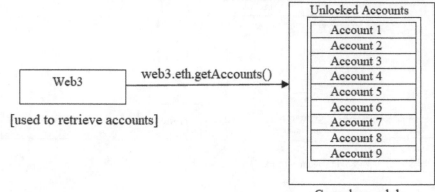

FIGURE 9.6   Fetching accounts from Ganache.

Ethereum Main network or Ethereum public test networks, the public and private keys are required to unlock the accounts, and the accounts should have some ether (wei) to deploy the contract.

The Main network uses real ether for deployment and manipulation or interaction with contract. Hence, the Ethereum test networks (Rinkeby, Kovan, Ropsten) which uses a fake ether, for development and testing purposes are generally used before deploying the contract to the Main network. The process for deploying the contract to the test network and Main network is the same.

9.5.2   FETCHING ACCOUNTS FROM GANACHE

It is very important to remember that we can deploy contract only when we have access to an account. Figure 9.6 depicts that web3 is used to retrieve unlocked accounts from genache module for deployment of smart contract.

The web3 library has many modules inside it, which are used to work with different types of networks. Here, "eth" is an Ethereum module in web3, which has a method "getAccounts()."

Code for fetching accounts from Ganache module

```
beforeEach(() => {
    // Get a list of all accounts
    web3.eth.getAccounts().then((fetchedAccounts) =>{
```

```
console.log(fetchedAccounts);
}); })
describe('contract_name', () => {
it('deploys a contract', () => {
console.log(accounts);
}); });
```

```
var firstAccount;

web3.eth.getAccounts().then((fetchedAccounts => {

firstAccount = fetched Accounts[0];

console.log("X: " + firstAccount);

}) console.log("Z: " + firstAccount);
```

FIGURE 9.7   Getting first account of Ganache module.

When we run this code in web3.
The console returns list of unlocked accounts fetched from Ganache module:

```
[ '0x992f437b0f15e1F789C8BD923b5C675e9c0c4u49',
'0x5738d9100A6CB64FB78730746ab61472c7808D9',
'0x7086D052EAaD559c7aCD8B993c6169aE0dC025',
'0x9fA18751b024FDC55cC85A484fF4261351Dd466',
'0x95a29de4A8cc4697E678622d270Aa74C2bd7494a',
'0xaED2F48c0d06a13D1FE2853E7Ee3B7d0aA32d3B',
'0x88A0941f0dcb501C20b7BaE5a41008EB981f1302',
'0xad5d1cB78e80518b596A814340826F36B89660Fc',
'0x67E67e936C9d22eDCf9ebC9A65720b313c54BC1' ]
```

But in order to get the first account of the array, in this case it's ['0x992f437b0f15e1F 789C8BD923b5C675e9c0c4u49'] the syntax illustrated in Figure 9.7 should be used.

9.5.3   Asynchronous Method for Fetching Accounts

Another method of fetching accounts from Ganache module is asynchronous is nature. Asynchronous action includes getting access to a contract, or a function in a contract or sending money from one person to another. Every function that we call with web3 is asynchronous in nature.

The "await" keyword is used since it is an asynchronous method.

```
//let accounts;
    beforeEach(async () => {
    // Get a list of all accounts
```

```
accounts = await web3.eth.getAcccounts();
});
describe('contract_name', () => {
it('deploys a contract', () => {
console.log(accounts); }); });
```

9.6 METAMASK EXTENSION

To deploy a smart contract, an account in Ethereum client like Metamask [15] or Mist Browser is required with some amount of ether. Metamask is a crome extension that permits common people to interconnect with Ethereum network. It can be used as Ethereum client. Metamask setup on the machine can be done by installing it from crome web store.

Metamask accounts have three distinct features associated with it. Figure 9.8 depicts the basic structure of Metamask with three distinct features.

Account address is something similar to username or email address that can be shared with anyone. Public key and private key together get combined to form password of the source and authorized the sending of funds from user's account to other account. All the three entities, namely account address, public key, and private key, are stored as hexadecimal number. When that hexadecimal number is copied to a JavaScript console, it automatically gets converted to base ten number, which is incomprehensively very large. This generated number uniquely constituent the private key. Private Key is not to be shared with anyone because if anyone ever gets to know it, then they can easily take all the funds that are in our contract.

Suppose we have to login into g-mail and yahoo mail, for the network we need different email accounts. But in Ethereum, one account is used for multiple networks

FIGURE 9.8 Metamask features.

that we ever interact with, i.e., one account can be created for all networks mainly Main, Rinkeby, Kovan, and Ropsten.

When an account is created in the Metamask, a 12-word mnemonic is displayed.

9.6.1 INFURA

Infura is a public API that gives us the ability to access a node on the network.

A 12-word mnemonic displayed after the creation of an account in Metamask is 12 random words which is used to generate (uses BIP39 algorithm) a series of accounts, each with a public key, private key, and an account address. So rather than memorizing public key, private key, and address of each account, we can just remember a 12-word mnemonic. The Rinkeby faucet (https://faucet.Rinkeby.io), shown in Figure 9.9, can be used to get some ether into the Metamask account.

The mnemonic is passed as an argument to the provider to unlock the accounts in the Metamask. To deploy the contract in public networks, a connection with a node on the public network is required. To become a node in Ethereum network, one needs to configure their machine, which is a little tedious job and requires a lot of time and effort. So, instead a service called Infura [16] is used to connect to a node already configured and existing on the network. Figure 9.10 shows the relation between Infura, provider, and account mnemonic.

The Infura has the nodes on all the Ethereum networks including the Main network as shown in Figure 9.10. Get the api key and the link to connect to the Rinkeby network from https://infura.io

As said above, a provider is needed to communicate with the Ethereum test network. The provider has to be setup manually specifying the network to connect and the account mnemonic to unlock the accounts. For this purpose, a node module "truffle-hdwallet-provider" should be installed and imported into the project. Truffle is a command line tool for contract creation. It is kind of one shot for development of Ethereum contract. Truffle is going to allow us to connect the Rinkeby network, illustrated in Figure 9.11, hosted through Infura and more importantly allow us to unlock and use accounts very easily.

As mentioned, when a smart contract is compiled by a solidity compiler, it generates a bytecode and ABI, which are required in the deployment of the contract on to

Rinkeby Ether Faucet

Give me your address and I'll give you .001 ether!

My Address: [   ] [Submit]

Great, coins are on the way!

If you're curious, here is your transaction id:
0x3a7e14dc1a1de12217577f385f336c2c8627795da5d63a74a81f4af04744a132

FIGURE 9.9   Rinkeby faucet.

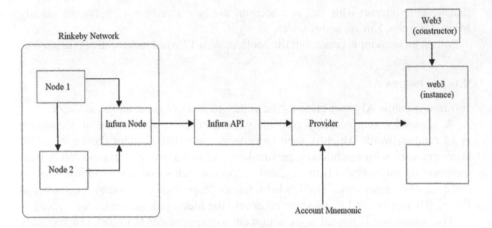

FIGURE 9.10 Connection through Infura Api.

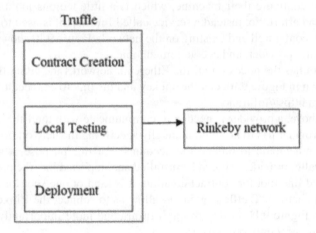

FIGURE 9.11 Connection with Rinkeby.

the Rinkeby network. To interact with the Ethereum network, the web3 instance is used. For example, the script to get access to the accounts in the Metamask is given below.

```
const accounts = await web3.eth.getAccounts();
```

The web3 library has many modules inside it which are used to work with different types of networks. Here, "eth" is an Ethereum module in web3, which has a method "getAccounts()." The await keyword is used since it is an asynchronous method. After getting the accounts, the contract can be deployed using one of the accounts. The contract deployment script is shown below.

```
// contract deployment script
   const result=await new web3.eth.Contract(JSON.parse(interface))
   .deploy({ data:'0x'+ bytecode,arguments:['']})
   .send({from: accounts[0],gas: '10000000'}); // gas limit
```

The "Contract" in the above code is a constructor in the Ethereum module for creating a new contract, which has "interface (ABI)" as an argument. The "deploy()" is a method in Contract which has "bytecode and arguments: [ ]" as the arguments. The arguments array contains all initial values or messages required for the contract. The "send()" method takes two arguments. One is the gas limit, and the other is the account address from which the contract is deployed. The send method will actually send this transaction to the network and the returns the address with many other parameters to the result variable. The contract is deployed to Rinkeby test network and the address value can be known from result.options.address.

Now the contract is deployed, a front-end app or a webpage should be created which can interact with the deployed contract. For this, the contract address and the ABI should be imported to the app. Whenever the Metamask is running inside the browser, it will automatically inject an instance of web3 library into the current active page. This web3 also has a provider which is connected to the public networks available in the Metamask.

Using this web3 with provider, contract address and the ABI form the contract, the app will be able to interact with the contract on the Rinkeby network.

9.7   TESTING WITH MOCHA FRAMEWORK

Whenever we compose a function for specific output, we can envision the functionality of that function and which parameters give what results. At the time of development of function, we can cross validate the function by running it and comparing the output with the expected one. This could be done by running the function in the console. In case, the output is not same as expected one then we can debug the code, run it again in the console, validate the result and keep on doing this till we get the expected or desired output.

But such "re-runs" are not perfect. To avoid such imperfect testing in solidity, we can use Mocha framework.

Mocha is a JavaScript testing framework for Node.js programs. Mocha is used for testing JavaScript code. Table 9.3 illustrates the basic function of Mocha framework.

We need to install Mocha framework in order to test Node.js programs.

Mocha testing framework can be installed by running following command in Node.js command prompt.

To install with npm globally, use the following code:

```
npm install --global mocha
```

TABLE 9.3
Mocha Functions

Function	Purpose
It	Run a test and make an assertion
describe	Groups together "it" functions
beforeEach	Execute some general setup code

or as a development dependency for respective project:

```
npm install –save-dev mocha
```

Mocha structure is illustrated in Figure 9.12. We deploy our contract in beforeEach block. "it" block is used for manipulating and making the assertion about the contract.

While writing tests in Mocha, we usually need an assertion library. We can choose any assertion library while working with Mocha framework. If we are using Mocha in a Node.js environment, then we can use the built-in assert module as our assertion library.

Testing of asynchronous code is very easy with Mocha testing framework. Generally, the following methods are used for testing asynchronous code with Mocha framework:

1. Testing with the help of a callback function.
2. Using promises (It is to be noted that all environments don't support promises).
3. Using async/await keyword (for environments that support async functions).

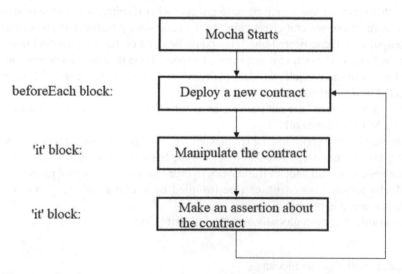

FIGURE 9.12 Mocha structure.

9.8   DESIGN OF A SMART CONTRACT

Now, let's design a smart contract. The name of the contract is Raffle.In this project, the complete money transactions is maintained and controlled by the smart contract. Firstly, we will design the layout of Raffle smart contract. Figure 9.13 shows the basic layout of the Raffle smart contract.

The created contract accepts ether from the users and marks them as contestants. And the contestants will transfer ether to contract, to enter in Raffle. The transferred ether will be held in the prize pool. Suppose two contestants, namely contestant 1(C-1) and contestant 2(C-2) participates in Raffle and each of them sends two ether, then prize pool will have total four ether.

A third-party agent that we called as an "executive" will tell the contract to pick a winner as shown in Figure 9.14. The contract will choose any one of the player and will send all of the money out of the prize pool to that particular winner.

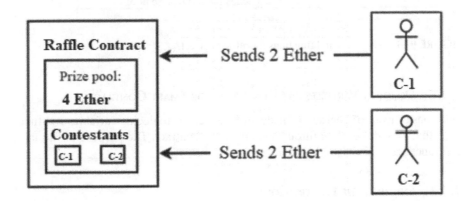

FIGURE 9.13   Raffle contract layout.

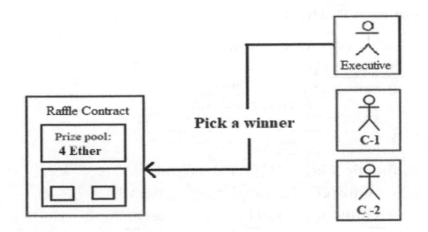

FIGURE 9.14   Function of executive.

Variables	Raffle Contract
Name	**Purpose**
executive	address of person who created the contract
contestants	array of addresses of people who have entered

Functions	
Name	**Purpose**
enter	enters a player into the lottery
pickWinner	Randomly picks a winner and sends them the prize pool

FIGURE 9.15   Variables and functions for raffle contract.

9.8.1 DECIDING VARIABLES AND FUNCTIONS FOR SMART CONTRACT

Once basic layout of contract is ready, the next task is to decide variables and functions that we will be using throughout our smart contract. The purpose of variables and functions is illustrated in Figure 9.15.

9.8.2 ENTERING THE LOTTERY CONTRACT

```
pragma solidity ^0.5.12;
contract Raffle {

    address public executive;
    address[] public contestants;
    function Raffle() public {

        executive = msg.sender;
    }

    function enter() public payable {
        require(msg.value > .01 ether);
        contestants.push(msg.sender);
    }
```

TABLE 9.4
Properties of "msg" Global Variable

Property Name	Property
msg.data	"Data" field from the call or transaction that invoked the current function
msg.gas	Amount of gas the current function invocation has available
msg.sender	Address of the account that begins the current function invocation
msg.value	Amount of ether (in wei) that was sent along with the function invocation

In above code, the function Raffle and enter has "msg" object. This object has some properties on it that describe both who just sent in a transaction to the network and some details about the transaction itself. We will discussed it in details.

Table 9.4 shows the property name and its associated functionality of the "msg" global variable.

It is significant to note that this msg global variable is available at both the time when we send a transaction in and when we do a call. So, anytime a function runs inside of our contract we will always have this msg object available.

9.8.3 VALIDATION WITH "REQUIRE" STATEMENT

Whenever someone calls the enter() function, contract takes their address and add it to array of contestants. Contestants array contains list of addresses.

The function enter() is of type payable, shown in Figure 9.16, because the person need to pay some ether to enter in the Raffle. "'public" is a keyword that indicates the access modifier or who can access this function.

The keyword "payable" is a modifier which indicates that this function accepts ether, as shown in Figure 9.16. So, when calling this function if the user has sent some ether, it will be directly added to the contract balance. Without the payable modifier the function will not accept the ether that has been sent. If it is then the function executes and he will be added as an approver or contributor. As mentioned above, approvers is a mapping between address and Boolean.

As shown in Figure 9.16, "require" function is used for validating the entry of contestants in the Raffle contract. The require will check the condition whether ether sent is greater than minimum required value to enter the contract. We can pass in some type of boolean expression in this "require" function. If that boolean expression returns false, then the entire function is immediately exited and no changes are made to the contract. If the expression evaluates to true then code inside the function continues to run usual. So, "require" function is used to make sure that some

```
function enter() public payable {
    require(msg.value > .01 ether);
    contestants.push(msg.sender); }
```

FIGURE 9.16 Require statement.

requirements has been satisfied or fulfilled before allowing all the other function to be executed.

The "msg.sender" will have the address of the person who invoked the function.

9.9   PSEUDO RANDOM NUMBER GENERATOR

As of now, contestants have the ability to enter into a raffle contract through the enter() method. The executive of the contract should then be able to call the pick winner function. The goal of this function is to randomly pick a winner out of a list of contestants. So, the function looks at the list of contestants that are in the contestant's array and randomly pick one entry in there and declares that contestant as a winner.

If we were using any other language besides solidity, let's say JavaScript, ruby, java to pick a winner. We might say just randomly pick a number and then using that number pick random element or a random index out of our contestants array.

However, with solidity we do not get access to a random number generator. Theirs is nothing like a random number generator with solidity. So now we're going to kind of fake the random aspect of this.

9.9.1   Methodology Used for Pseudo Random Number Generator

We're going to take the current block difficulty, current time and the list of addresses of all the people who have entered into our contract. Current block difficulty takes some amount of time to process an actual transaction, the amount of time that it takes to pick or solve a block or closeout a block is referred to as the block difficulty. This is represented as an integer.

These three inputs and feed them into the SHA3 algorithm, and it spits out a really big number in hexadecimal, illustrated in Figure 9.17. The generated big hexadecimal number will be used to pick our random winner.

The code shown in Figure 9.18 to randomly pick a winner is what we refer to as a pseudo random number generator.

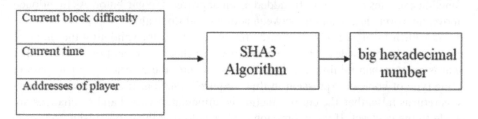

FIGURE 9.17   Random number generation.

```
function random() private view returns (uint) {
    return uint(SHA3(block.difficulty, now, contestants));
}
```

FIGURE 9.18   Code for a pseudo random number generator.

The function name is "random" because its goal is to generate a random number. This is a "private" function because we really don't want anyone else to write this or call this function; it's just the developer that defines this function. The function is marked as a "view" type because we are not modifying any state or any data in the contract. The only goal of this function is to return a random number.

The block variable is a global variable that we have access to at any given time. Difficulty will be a number that indicates how challenging it is going to be to seal or solve the current block.

Now inside the random function we do have to make sure that we return an unsigned integer but a hashing algorithm that returns a hash. So, we can convert this hash to an integer.

9.9.2 SELECTING THE WINNER

The random number generated above can be used to pick any player randomly from the contestants array. Figure 9.19 depicts the methodology used for selecting the winner of Raffle.

Property of modulo operator can be used to get the index of the winner inside our contestants array. So, random modulo contestants.length (this is the number of contestants who have entered into the Raffle) will return the remainder of division between random and contestants.length. This number (remainder) is always going to be between 0 and contestants.length. So, it can be used as the as the index of the winner inside the contestants array.

Function for selecting the winner from contestants array is shown in Figure 9.20.

9.9.3 SENDING ETHER FROM CONTRACT

Now, we have the index of the contestant who won the Raffle so we can select that contestant out of contestants array and then send money from our contract.

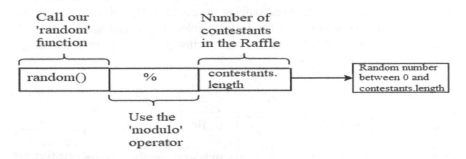

FIGURE 9.19 Methodology for selecting the winner.

```
function pickWinner() public restricted {

    uint index = random() % contestants.length; }
```

FIGURE 9.20 Code for selecting the winner.

⇨ `contestants[index].transfer(this.balance);`

We can send prize pool money to the winner address by calling.transfer function. This function is going to take all the money collected in prize pool and then send it to the winner's address that is specified in [index]. So, this is a function that is available on every address that we store inside of solidity.

Working of transfer function

The transfer function will attempt to take money from the current contract and send it to winner address.

⇨ `contestants[index].transfer (1)`

The above line of code would transfer 1 wei to winner address.

9.9.4 RESETTING CONTRACT STATE

Now, we are going to talk about one of the big issues around our current contract at present. After picking a winner, we send someone some amount of money. In an ideal world, we might try to reset our contract so we can run another round of the Raffle immediately after picking a winner. In other words, it'd be really nice if we could just have an infinite series of Raffle contract just running one after another without redeploying the contract.

Figure 9.21 helps us to understand how to approach the above stated problem.

When our contract is first created, we have an empty player's array. Now, consider that one or two people might enter the contract or the lottery and their addresses are added to the player's array.

Now here's the interesting part.

When we pick a winner if we empty out our list of addresses, we can then essentially reset the state of our contract back to how it was first deployed. So by emptying out the list of players, we can reset the state of our contract and automatically get it ready for another round without redeploying the contract. By doing so, we can run the contract infinitely without ever having to redeploy it again and again.

Figure 9.22 shows updated pickWinner () function for resetting contract state.

⇨ `players = new address[] (0);`

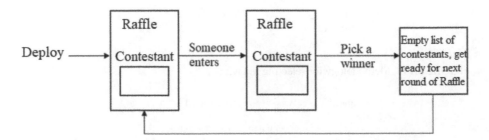

FIGURE 9.21   Resetting contract state.

```
function pickWinner() public restricted {

    uint index = random() % contestants.length;

    contestants[index].transfer(this.balance);

    contestants = new address[](0); }
```

FIGURE 9.22   Updated pickWinner() function.

Above line of code creates a new dynamic array of type address. "(0)" marks that we want to create a new array of type address. We want that array to be dynamic in nature and have an initial size of 0.

9.10   REQUIRING MANAGERS

Our random function is pseudo random in nature. So, if some stranger realized that if they called "pickWinner" function at the right time, they would potentially make sure that a given player won the lottery.

So, in short we want to make sure that solely the executive is able to call that pickWinner function.

Inside the pickWinner function, we need to update the code shown below.

```
modifier restricted() {
   require(msg.sender == executive);
   _;
}
```

9.11   SMART CONTRACT FILES

A final smart contract consist of many files associated with its deployment, testing, compilation, file containing installed dependencies, etc. For Lottery contract, we have five files, namely Lottery.sol file, this file contains set of rules/logic to handle the state of the asset stored in the smart contract. Then, we have Lottery.test.js file, this file contains all the test cases for our smart contract. Another files are package.json file, compile.js file, and deploy.js file. Detailed view of all these files is given in further section.

9.11.1   PACKAGE.JSON FILE

This file shows the complete information about the project, the installed dependencies, the start scripts, test scripts, and build scripts. This file gives an idea about the project and how to execute the project.

```
1 ▾ {
2      "name": "raffle",
3
4      "version": "1.0.0",
5
6      "description": "",
7
8      "main": "index.js",
9
10 ▾   "scripts": {
11        "test": "mocha"
12      },
13
14      "license": "ISC",
15
16 ▾   "dependencies": {
17        "ganache-cli": "^6.6.0",
18        "mocha": "^6.2.0",
19        "solc": "^0.4.17",
20        "truffle-hdwallet-provider": "0.0.3",
21        "web3": "^1.0.0-beta.26"
22      }
23 }
```

9.11.2 RAFFLE.SOL FILE

This file contains the smart contract code written in solidity language. All the variables and functions discussed in Section 9.8.1 is used in this. This file gives an idea how to write the smart contract.

```
1  pragma solidity ^0.5.2;
2 ▾ contract Raffle {
3      address public executive;
4      address[] public contestants;
5 ▾    function Raffle() public {
6          executive = msg.sender;
7      }
8 ▾    function enter() public payable {
9          require(msg.value > .01 ether);
10         contestants.push(msg.sender);
11     }
12 ▾   function random() private view returns (uint) {
13         return uint(keccak256(block.difficulty, now, contestants));
14     }
15 ▾   function pickWinner() public restricted {
16         uint index = random() % contestants.length;
17         contestants[index].transfer(this.balance);
18         contestants = new address[](0);
19     }
20 ▾   modifier restricted() {
21         require(msg.sender == executive)
22         _;
23     }
24 ▾   function getContestants() public view returns (address[]){
25         return contestants;
26     }}
```

9.11.3   RAFFLE.TEST.JS FILE

```
1   const assert = require('assert');
2   const ganache = require('ganache-cli');
3   const Web3 = require('web3');
4   const web3 = new Web3(ganache.provider());
5   const { interface, bytecode } = require('../compile');
6
7   let raffle;
8   let accounts;
9   beforeEach(async () => {
10      raffle = await new web3.eth.Contract(JSON.parse(interface))
11      .deploy({ data: bytecode })
12      .send({from: accounts[0], gas: '10000000'});
13  });
14
15  describe(' Raffle Contract', () => {
16      it('deploys a contract', () => {
17          assert.ok(raffle.options.address);
18      });
19
20      it('allows multiple accounts to enter into the contract', async () => {
21          await raffle.methods.enter().send({
22              from: accounts[0],
23              value: web3.utils.toWei('0.05', 'ether')
24          });
25
26          await raffle.methods.enter().send({
27              from: accounts[1],
28              value: web3.utils.toWei('0.02', 'ether')
29          });
```

```
30          await raffle.methods.enter().send({
31          from: accounts[2],
32          value: web3.utils.toWei('0.02', 'ether')
33  });
34          const contestants = await raffle.methods.getContestants().call({
35              from: accounts[0]
36          });
37          assert.equal(accounts[0], contestants[0]);
38          assert.equal(accounts[1], contestants[1]);
39          assert.equal(accounts[2], contestants[2]);
40          assert.equal(3, contestants.length);
41          });
42      it('minimum amount of ether is required to enter the raffle', async () => {
43          try {
44          await raffle.methods.enter().send({
45              from: accounts[0],
46              value: 0
47          });
48          assert(false);
49          } catch (err) {
50          assert(err);
51          }
52          });
53      it('only executive can call the pickWinner' , async () => {
54          try {
55              await raffle.methods.pickWinner().send({
56                  from: accounts[1],
57              });
```

```
58        assert(false);
59 ▾    } catch (err) {
60          assert(err);
61      }
62   });
63 ▾ it('sending money to the winner and resets the contestants array',async () => {
64 ▾   await contestants.methods.enter().send({
65        from: accounts[0],
66        value: web3.utils.toWei('2', 'ether')
67
68      });
69
70      const initialBalance = await web3.eth.getBalance(accounts[0]);
71
72      await raffle.methods.pickWinner().send({from: accounts[0] });
73
74      const finalBalance = await web3.eth.getBalance(accounts[0]);
75
76      const difference = finalBalance - initialBalance;
77      console.log(finalBalance - initialBalance);
78      assert(difference > web3.utils.toWei('1.8', 'ether'));
79    });
80  });
81
```

9.11.4   COMPILE.JS FILE

```
1  const path = require('path');
2  const fs = require('fs');
3  const solc = require('solc');
4  const lotteryPath = path.resolve(__dirname, 'contracts', 'Raffle.sol');
5  const source = fs.readFileSync(rafflePath, 'utf8');
6  module.exports = solc.compile(source,1).contracts[':Raffle'];
7
```

9.11.5   DEPLOY.JS FILE

```
1  const HDWalletProvider = require('truffle-hdwallet-provider');
2  const Web3 = require('web3');
3  const { interface, bytecode} = require('./compile');
4
5  const provider =  new HDWalletProvider(
6    'payment hen clump pave scatter august satisfy swear scheme spread saddle fame',
7     'https://rinkeby.infura.io/v3/aa4a1902bc154228acfd09e0eec205cc'
8  );
9  const web3 = new Web3(provider);
10
11 ▾ const deploy = async () => {
12    const accounts = await web3.eth.getAccounts();
13
14    console.log('Attempt to deploy from account', accounts[0]);
15
16  const result = await new web3.eth.Contract(JSON.parse(interface))
17    .deploy({data : bytecode})
18    .send ({ gas: '1000000', from: accounts[0]});
19
20  console.log('Contract deployed to', result.options.address);
21  };
22  deploy();
```

9.11.6   RUNNING SMART CONTRACT

```
C:\Users\sharm\lottery>npm run test
  >inbox@1.0.0 test C:\Users\sharm\lottery
  > mocha
  Lottery Contract
  (node:15376)
  √ deploys a contract
  √ allows multiple accounts to enter (612ms)
  √ require a minimum amount of ether to enter (74ms)
  √ only executive can call the pickWinner (108ms)
1999958005999992800
  √ sends money to the winner and resets the contestants array (279ms)
  5 passing (2s)
```

Run the command "npm run test" in Node.js command prompt.

CONCLUSION

The main goal of this chapter is to give a comprehensive approach on how to write smart contracts. In this chapter, a brief description about the various testing framework, libraries, and how it works has been given lucidly. Further, this chapter focuses on the deployment of the contract in Ethereum Rinkeby test network. Smart contract, which is one of the key functions of Ethereum Dapps, has been explained in an elaborate way. Then this chapter mainly focuses on how various functions are decided while designing a smart contract. This chapter talks about the whole development, deployment, and interaction part in a profound manner. Using the blockchain technology, the security, privacy, and the trust can be enhanced exponentially. Blockchain is a disrupting technology, and in the very future, it will change whole economic and commerce systems.

REFERENCES

1. V. Buterin, A next-generation smart contract and decentralized application platform. https://github.com/ethereum/wiki/wiki/White-Paper/, 2019.
2. S. R. Niya, F. Shüpfer, T. Bocek, and B. Stiller, Setting up flexible and light weight trading with enhanced user privacy using smart contracts, NOMS 2018-2018 IEEE/IFIP Network Operations and Management Symposium, Taipei, pp. 1–2, 2018.
3. K. Zhang and H. Jacobsen, Towards dependable, scalable, and pervasive distributed ledgers with blockchains, 2018 IEEE 38th International Conference on Distributed Computing Systems (ICDCS), Vienna, pp. 1337–1346, 2018.
4. S. Nakamoto, Bitcoin: A peer-to-peer electronic cash system. http://bitcoin.org, October 2008.
5. F. Hofmann, S. Wurster, E. Ron, and M. Böhmecke-Schwafert, The immutability concept of blockchains and benefits of early standardization, 2017 ITU Kaleidoscope: Challenges for a Data-Driven Society (ITU K), Nanjing, 2017, pp. 1–8, 2017.

6. A. Anoaica and H. Levard. Quantitative description of internal activity on the ethereum public blockchain. In 2018 9th IFIP International Conference on New Technologies, Mobility and Security (NTMS), pp. 1–5, Feb 2018.

7. T. Chen et al., Understanding ethereum via graph analysis, IEEE INFOCOM 2018-IEEE Conference on Computer Communications, Honolulu, HI, pp. 1484–1492, 2018.

8. Y. Chen, S. Chen, and I. Lin, Blockchain based smart contract for bidding system, 2018 IEEE International Conference on Applied System Invention (ICASI), Chiba, pp. 208–211, 2018.

9. G. Maxwell, "Coinjoin: Bitcoin privacy for the real world," in Post on Bitcoin Forum, 2013.

10. A. Aldweesh, M. Alharby, E. Solaiman, and A. van Moorsel, Performance benchmarking of smart contracts to assess miner incentives in ethereum, 2018 14th European Dependable Computing Conference (EDCC), Iaşi, Romania, pp. 144–149, 2018.

11. J. S. Coron, "What is cryptography?" *IEEE Security &Privacy Journal*, 12(8), 70–73, 2006.

12. B. K. Mohanta, S. S. Panda, and D. Jena, An overview of smart contract and use cases in blockchain technology, 2018 9th International Conference on Computing, Communication and Networking Technologies (ICCCNT), Bangalore, pp. 1–4, 2018.

13. H. Cui-hong, Research on Web3.0 application in the resources integration portal, 2012 Second International Conference on Business Computing and Global Informatization, Shanghai, pp. 728–730, 2012.

14. A. Manzalini and A. Stavdas, A service and knowledge ecosystem for Telco3.0-Web3.0 applications, 2008 Third International Conference on Internet and Web Applications and Services, Athens, pp. 325–329, 2008.

15. G. Foroglou and A.-L. Tsilidou, "Further applications of the blockchain," 2015.

16. D. Kraft, "Difficulty control for blockchain-based consensus systems," *Peer-to-Peer Networking and Applications*, 9(2), 397–413, 2016.

10 Blockchain in Healthcare Sector

S. Porkodi and D. Kesavaraja
Dr. Sivanthi Aditanar College of Engineering

CONTENTS

10.1 INTRODUCTION

10.1.1 TODAY AND TOMORROW

In the new world of digital era, technologies play important roles in the day-to-day life of the humans. Every industry tries to adapt the new technology that can be used in a much useful and productive way. One such new technology is blockchain. That too in healthcare industry, blockchain can be used to solve many problems in the healthcare industry such as medical data security, accessibility, interoperability, etc. In future, blockchain has the potential to help authentic, personalized, and secure healthcare systems by merging patient real-time health data and updating healthcare details to the patient and can yield greater success in the medical field.

10.1.2   VISION

The vision is to adapt blockchain technology (BCT) in the healthcare centers and hospitals to get maximized benefits. Using BCT could lead to a betterment in healthcare systems such as secure data transaction, protecting user details, efficient data management, right treatments can be given to patients at right time by efficient diagnosis of patient's health. From government hospitals to smaller clinical centers can adopt the BCT for a higher potential treatment.

10.1.3   HEALTHCARE

Health care is a data-oriented industry, where very vast data are being created, accessed, and then disseminated regularly. It is crucial to store or disseminate these huge sets of data. It is a significant challenge to handle the data, as these data are highly sensitive and have some limiting factors like privacy and security [24,36].

In the smaller clinics as well as small healthcare field scalable, secure, and safe sharing of data is much important for the combined decision making and efficient diagnosis. Sharing data is essential and important to transfer medical data reports from the physicians or doctors to the consulting patients for quick update on the progress of the treatment. The physicians and doctors transfer the data or report to their patients with high privacy on a regular basis, in order to make sure that both sides consist of updated information regarding the health condition of the patient. In the endlessly utilized huge area big domains such as telemedicine and e-health record (EHR) in which the clinical reports are sent remotely to a distant location specialist for consulting the opinion of expert. In these types of systems, the data of the patient is sent via real-time online monitoring (such as telemetry or telemonitoring, etc.) or storing and forwarding technology [3,8,22,29]. By using these online technologies, the remote diagnosis of the patients can be carried out and treated by experts with the help of transferred clinical reports. In these technologies, the privacy, security, and sensitivity of the clinical reports are the main challenges which occur since the data of the patients are sensitive in nature. Thus transferring of data in a scalable, secure, and safe way is very much important to support a meaningful and healthy communication between the remote patients and the experts. This successful safe transfer of clinical reports yields related confirmations or recommendations from the clinical experts or specialists that could be resulted in improvement in the accuracy of diagnosis and a much effective treatment [7,9,13,16,65].

10.1.4   CHALLENGES IN HEALTHCARE

In the field of healthcare, there are different challenges in interoperability. For example, the successful, safe, and secure transactions of clinical report data among the research institutes or healthcare organizations can have lots of challenges, when applied in practical operations. These clinical report data needs to be trustworthy, substantial, and mainly it should be a healthy collaboration among the involved entities. Some of the potential constraints that this industry faces are sensitivity of data, governing rules, data sharing procedures and agreements, ethical politics, and patient matching complex algorithms. There are few important terms, which have to be mutually agreed before the clinical data transaction process [17,50]. In the near past, the researchers were trying to develop and implement applications based on Internet of Things (IoT), Computer Vision, machine learning (ML) and Artificial Intelligence (AI) to facilitate physicians and doctors in diagnosing the treatment with different chronic diseases.

10.1.5 BLOCKCHAIN

BCT acts as human memory, which secures data and keeps record for every data transaction among the users. Blockchain was first used in Bitcoin, and in the recent past, there are lots of interest in using the blockchain-based application for secure and safe data delivery in many fields such as healthcare data [4, 28,64], e-healthcare data sharing [2,60], biomedical [38], thinking and brain simulation. The blockchain uses peer-to-peer network, to obtain the distributive nature in public digital ledger. It consists of cryptography for protecting the data blocks of the user during transaction. Blockchain is designed to solve the limitations of synchronization in the distributive traditional database with the help of algorithm of distributive consensus. The basic characteristics of the BCT are open source, anonymity, transparent, decentralized, immutable, and autonomy as shown in Table 10.1.

People who involve in the blockchain network verify every new transaction that has been carried out. All the transaction blocks are verified by each network nodes and turn into immutable. The workflow process of the blockchain is shown in Figure 10.1.

In the near future, BCT would give potential help in reliable, secure, and personalized healthcare system by merging present up-to-date data of healthcare system

TABLE 10.1
Characteristics of Blockchain and Its Description

Characteristics	Description
Open source	The blockchain is an open-source network that provides access for all the people in connection to network. This feature not only allows anyone to check public records but also allows to develop and design various applications based on blockchain.
Anonymity	When a data is transacted from one node to another node, identity and details of both the parties remains anonymous. Hence, the blockchain becomes more reliable and secure system.
Transparent	The stored and recorded data present in blockchain are transparent to capable users. This property helps to prevent the alteration of data or data being stolen.
Decentralization	The data in the database can be monitored, accessed, updated, and stored on many different systems at the same time. There is no single centralized authority to manage data instead it has open access for anyone who are connected to network.
Immutable	Once the record is stored in the blockchain, it stays there forever, they can't be easily modified without the control of concurrent nodes which is more than half of nodes.
Autonomy	Blockchain is autonomous and independent, which means that any node in blockchain network can transfer, access, update, or store data in a safe and secure way. Thus, it makes the system free of external and intervention more trustworthy.

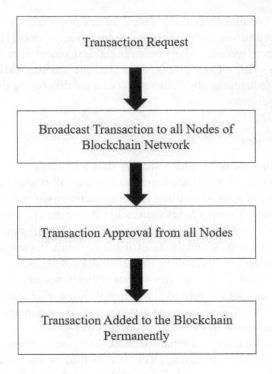

FIGURE 10.1 Generalized blockchain workflow.

along with real-time entire clinical report data of patient health. In this chapter, existing progress in blockchain and latest development of BCT in healthcare industry model are discussed.

10.2 RELATED WORKS

The potential of BCT to share data in the healthcare systems and to give assistance to many different diagnosis applications have been described already in many studies.

For example, private blockchain is used for storing and monitoring the clinical report data associated with method called Healthcare Data Gateway (HDG), which is developed by Yueet al. [63]. This system is a personalized health care in which the patients have freedom to monitor, access, and manage their healthcare summary and own clinical report data that has been stored in private blockchain network, which is a centralized database that gives access to only specific authorized users.

Chen et al. [14] studied that cloud storage integrated with blockchain framework can share and manage patients, own clinical data. This can be used for secure and safe storage of clinical data and patients, own clinical data can be exchanged. This is a unique approach where no third party is involved and only the patients have access to their own clinical data. Cryan [15] proposed blockchain-based innovative and systemic architecture in order to give protection to the sensitive personal data of the patients, solves issues of security to critical data, and blockchain-based software system is implemented throughout the hospital. Now a days, blockchain is considered as mechanism, security, privacy and share medical data are improved among healthcare entities and clinical experts or specialists.

Ivan [33] proposed a demonstration in which public blockchain is implemented for encryption of data, to store healthcare data safely. The blockchain network is a decentralized system database consists of open access to all the users who have connection with the network. The health data is encrypted and publicly stored which leads to a development of Personal Health Record (PHR) based on blockchain. This system allows the patients to access the clinical report data better; the patient can monitor, access, and contribute to their own files and also can share them with any relevant caregiving or healthcare agencies or experts.

Griggs et al. [25] studied that a private blockchain is adopted which is Ethereum protocol based in order to facilitate secure and safe usage of the biosensors and to eradicate risks in security regarding monitoring system of remote patients. In this strategy, real-time secure remote monitoring can be done, thus physicians and doctors are allowed to keep their patients' health status updated even from a far location and can also maintain secure, safe, and updated patient history. Jiang et al. [34] designed unique healthcare system for information exchange based on blockchain called BloCHIE. This system evaluates the requirements to share the healthcare clinical data for electronic and personal clinical data records, also deals with different forms of clinical data as the data is generated from different sources that is collected in the blockchain. Privacy and authenticity satisfactory requirements are validated by coupling platform along with off-chain and on-chain processes of verification.

In recent past, Wang et al. developed an artificial parallel execution healthcare system based on blockchain framework [59] to maintain and evaluate the health status of diseases of the patient. This method also evaluates overall status, diagnosis, treatment, patient condition, and the related therapeutic steps and process via parallel execution analyses and computational trials in order to make correct clinical decisions. This system is tested and evaluated to find the effectiveness in the treatment and accuracy of the diagnosis.

The research of blockchain reaches high potential in clinical and biomedical domains. The BCT can be utilized for storing all the clinical data plans, protocol, and consents even before the commencement of clinical examination or trails. Thus, sensitive data that are relevant to the trail are up to date already, transparent to public, time stamped and secure. Smart contracts can be executed, replicated, and deployed in different phases of trails for ensuring the transparency.

Shubbar [52] proposed a method based on blockchain called DermoNet, to assist patients of dermatology through the online consultation of dermatologist via a tele-dermatology monitoring system. Shubbar [52] suggested a telemonitoring healthcare based on blockchain framework to diagnosis and treat the cancer tumors for patients in remote location. Smart contracts are used, which are extensively used in order to validate and maintain security to patient's clinical data at the specialized clinical centers and impatient homes.

ProActive aging [31] is a platform based on the BCT that is designed to support aging people in active living. In the processes of extensive clinical treatments like

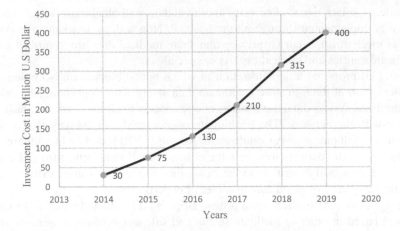

FIGURE 10.2   Worldwide blockchain capital market investment (2014–2019).

aging, surgical operations, and chronic diseases like cancer, the blockchain could be well suited and ideal choice. The biomedical researchers, drug manufactures, and pharmaceutical industries can use the DNA data that is stored in blockchain to do advance researches at level of genomic.

In the current global digital era, the growth of the blockchain can be determined with the growth in communication and information technologies. Based on a statistical survey that was taken in 2014 by Statista, an observation of vast increase in blockchain investment and funding can be seen worldwide as shown in Figure 10.2. Thus, the rate of expansion and attaining new avenues in the blockchain progress are highly expected.

10.3   APPLICATION OF BCT IN HEALTHCARE

Initially, BCT has been developed for Bitcoin which is a well-known cryptocurrency; however, in current world, blockchain is used in many fields and industries including health care and biomedical [38].

BCT has high potential which can be seen in some medical fields such as EHR system, medical fraud deduction system, clinical researches, pharmaceutical industry researches, and neuroscience by its nature to secure and stabilize dataset by which the patients, doctors, researchers, or any other users interact via different transactions as shown in Figure 10.3.

10.3.1   E-Health Record System

In the past few years, there is an increased need for digitalized medical or clinical health data and records. This was caused due to the hospitals, healthcare devices, physicians, and doctors since the digitalized data produces easy sharing and access to the data and also fast and better decisions can be made. EHR system is one of the most important and common application in healthcare industry based on blockchain as shown in Figure 10.4.

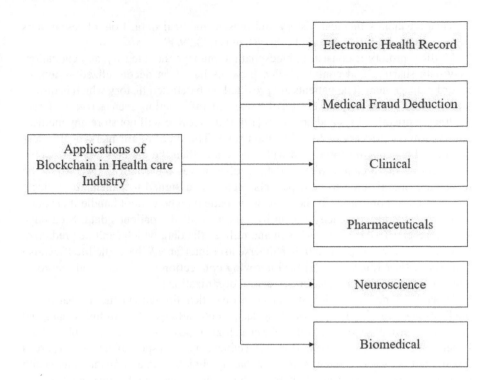

FIGURE 10.3 Applications of blockchain technology in healthcare.

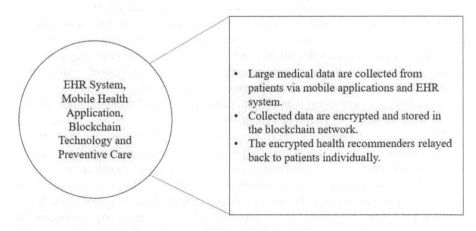

FIGURE 10.4 Description of interaction between the elements of EHR system.

There is no way to handle the lifetime clinical records in the EHR system that has been used in multiple different hospitals or health cares as these situations usually separate data from one provider to another. Due to these problems, access to the medical history and past data are lost. So there arises a need for an innovative way, in order to handle the data of EHR systems. By which patients are encouraged to engage patient's medical history and present medical data. Lots of researchers propose blockchain to maintain EHR system [11,23,39,49].

MedRec prototype consists of blockchain to manage the integrity, authentication, easy data sharing, and confidentiality. It works based on decentralized system of record management. The patients are provided with detailed history which is immutable and provides easy access for the relevant information even across different healthcare organizations and providers [19]. But MedRec will not store any medical data or any adjustment and modification time. The records are marked in blockchain, and patients are forwarded with advice and then the patients are responsible for selecting the location to store data. It also moves the control power from the healthcare organization to their patients, weight is assigned to patients to indicate the owner's responsibility. If there are some patients who cannot handle their data, specific agents are assigned to them by admin to fill the patient's data. So a huge size of patients individually enters in and utilizes the data which can't be predicted, they can create work, plan, and then diverse user interface. Where, the MedRec also consists of User Interface (UI) for improving connection to the data and record of health care, which circulates across various organizations.

When implementing EHR system, there arise critical limitation issues in sharing the medical data such as auditing, controlling data, secure tailing of data in clinical data, and data provenance. Due to these issues, Xia et al. [62] proposed secure and safe blockchain system called MeDShare in order to perform clinical data transportation between parties that are untrusted. MeDShare is used for sharing clinical data and maintaining health data and records in electronic form among hospitals, providers who give cloud service and researchers of healthcare field with audit control in a personalized form, better data provenance, and limited possibility of threats for data privacy and security.

EHR system usually consists of critical and high sensitivity that are data relevant to patients and shared frequently among healthcare providers, clinicians, pharmacists, researchers, and radiologists for providing effective and easy diagnosis and treatment. When these high-sensitive data of the patient are distributed, transmitted, or stored across different entities, the treatment of the patient could be compromised that would pose a huge threat to the health of the patient and maintain updated history of the patient. For example, if a patient is fighting against chronic diseases such as HIV or cancer, commonness of risk becomes much high along with pre-treatment and post-treatment history and rehabilitation steps or followups. Hence, updated history of patient is necessary to be maintained to carryout treatment effectively. Due to these issues, Dubovitskaya et al. [18] developed a framework based on blockchain to share, maintain, and manage the cancer patient's clinical records in electronic form. A blockchain is adopted with permissions for storing, managing, and accessing the patient's information in an encrypted form. This type of frameworks are used for practical implementation of blockchain to access the patient data, manage the security and privacy of patient clinical data history, and current data in practices.

In 2016, project such as Estonian medical record based on blockchain is a benchmark in the medical history, where Estonia is merged along with blockchain's global leadership. It is proposed with an idea to keep more than millions of private clinical records and also for the wide availability of the records for insurance companies [27] and medical providers.

The reason for the growth of blockchain at a global level in medicine may be that it has strong patient assurance. By using the blockchain, the user's health records can be made unaltered and immutable. If there is any adjustment made or unauthorized access can easily be recognized anywhere in blockchain. It is used for maintaining data integrity, identify criminal activities like records of adulteration or wholesale fraud, review is made simpler, and sharing medicinal record can be approved. When a patient visits, it can be noted by majority of suppliers of the patient quickly. Drug solution, hypersensitivities, and medication bugs are accommodated by patient caring algorithm throughout blockchain very quickly even without the requirement of any difficult forms of pharmaceutical compromise. The innovation of BCT is encouraged to improve access to healthcare data, encourage confirmation of medical data, managing clinical record [43,45], high effective care arrangement, and expanded security.

10.3.2 MEDICAL FRAUD DEDUCTION SYSTEM

One of the important huge blockchain applications in the medical field is the management of supply chain which consists of medical drugs. The supply chain management in the health care is more important due to its highly growing complexity. As if there occurs any compromise in supply chain of healthcare industry, in turn, it would greatly affect the patient's wellbeing [35]. These supply chains of medical drugs are vulnerable. It contains holes in which any fraudulent attacks may happen since it consists of many moving peoples and parts. The usage of blockchain enables a secure and safe platform by eliminating the issues such as traceability of product is improved, higher-order data transparency is introduced and prevents the occurrence of fraud. A blockchain record undergoes process of validation and updated using smart contracts. So it is not easy to manipulate a blockchain [41].

10.3.3 CLINICAL RESEARCH

The issues that may occur in the clinical trials include issues, such as data enrolling [30], data privacy, record management, data sharing, and data integrity. The researchers of the healthcare field work on blockchain to resolve the issues that occur during the clinical trials by using blockchain [44]. In future, there could be a storm of applications based on blockchain, machine learning, and AI.

Nugent et al. [46] proposed a study in which functionalities of smart contracts in the blockchain [12,61] are provided by a protocol called permissioned Ethereum that is used in parallel with the medical information management system. The problem of patient enrollment issue is discussed in this work. The study shows that the Ethereum results in much faster transactions when compared with Bitcoin, thus the conclusion suggests to use smart contracts in Ethereum to provide data management and transparency in the clinical trials. The blockchain-based patient enrollment belongs to the existing application in clinical research. Benchoufi [5] research is based on a blockchain framework, which is implemented in order to obtain all patient's contents that can be stored and used for tracking in a secure way, untestability, and verifiable.

10.3.4 PHARMACEUTICAL INDUSTRY RESEARCH

In the healthcare industry, one of the leading sectors and largest growing industry is pharmaceutical. Basically, the pharmaceutical industry helps to introduce potential and new drugs in markets. When the drugs are sold to patients its responsibility is to determine validity information for the medical products, to ensure safety, to evaluate [43] and process to produce safe drugs for patients so that they can recover quickly. The usual challenges faced by the drug manufacturing company is that they cannot timely track their own products, which may create many risk and issues such as invading fake drugs to circulate in use, allow counterfeiters in compromising production. One of the healthcare industry risks that has been observed throughout the world is producing and distributing the counterfeit drugs that too majorly in the developing and under-developed countries. Blockchain can be best-used technology in the times of Research and Development (R&D) and production of drugs. Blockchain are used to ensure potential drug's production steps, monitor, and evaluate. In the recent past, a project of blockchain-based counterfeit medicine is launched by research foundation called Hyperledger [55,58]. It was launched to inspect and fight against the production and creation of the counterfeit drugs. The drugs are needed to be evaluated, monitored, and ensured that digital technologies are used in all the processes of development and supply in the drugs of pharmaceuticals for effective delivery in authenticating drugs for patients and reliability throughout the world that too mainly in the developing and under-developed countries. Digital Drug Control System (DDCS) is used to solve these issues of counterfeit drugs [48]. Some larger industries in the pharmaceutical (such as Amgen, Pfizer, and Sanofi) joined together and launched a pilot project DDCS based on blockchain to inspect and evaluate the new drugs that are released in the global market. The blockchain-based system tracks the drug location at any time, track the production of drugs [40], falsified drug traceability [54] is improved, and guarantee could be given for the drug quality supplied to the patients [48] and the supply system of the drugs [56] are secured.

10.3.5 NEUROSCIENCE

In the healthcare industry, one of the growing discipline is neuroscience. The neuroscience is developing in a new style by excluding all their mechanical interactions with its infrastructure of the surrounding and also allowing a user to control data and devices via mental commands and thought wave. These neural devices have the capability for interpreting the brain activity patterns and translate the patterns to commands which can be used to control the external devices, and based on the brain activity patterns, the present person's mental state can be easily deducted. The devices such as 'Cranial Dopplers', 'Electromyographs', 'Implantable Neurostimulators', and 'Neurosurgery Surgical Robots' that consist of neural interface consist of special task such as they can read and interpret the signals of brain. These devices have wireless communication, chips for computing, and sensitive sensors. Electrical signals regarding the brain activity are read, deciphered, and the signals are transmitted.

All these activities occur in the single wearable device on the user head. Blockchain is used to record all the brain signals and thought waves in neural interface where the big data and complex algorithms can utilize the signal data [20,53].

In 2017, a company named Neurogress located in Geneva confirms using BCT. This company aims to build systems based on neural control, which gives access for users to control drones, Virtual Reality (VR)/Augmented Reality (AR), smart applications, and even robotic arms only with the users own thought waves and brain signals. The control systems of the Neurogress need brain data more than 90% for training AI systems based on machine learning algorithms for improving the accuracy of brain reading technique. The company released a white paper demanding a neural activity of the user as big data by citing Human Brain Project. The project demands exabytes of memory (1 billion gigabytes = 1 exabyte) because the samples used need such capacity to store data. Thus, the company Neurogress prefers the BCT to handle such big data, along with privacy and security, and handle the storage capacity problem effectively. So the data that is been recorded on the decentralized blockchain is resistant to attacks such as hacking and thus become more private and secure. The blockchain usage can also make the system of Neurogress, transparent and open to the users who use the platform services of Neurogress, but the system has the capacity to easily rack any such abnormal activities that would occur in the system ensuring user data confidentiality and security.

The blockchain is basically an innovation in the information technology field that could lead to the development of more important applications of neuroscience department in the future such as brain thinking, brain stimulation, and brain augmentation. The blockchain is mainly used to store the digital version of the whole human brain. The mind files are stored to act as information building blocks in the chain of personal thinking which can be shared in the network of peer-to-peer file system to do historical versioning. The computational input – processing – output system is proposed based on blockchain thinking, that consist of many features which could provide potential integration, human enhancement, and AI. The computers are interconnected in the network of blockchain, the systems can perform handshake during the timestamp intervals in order to validate source and ledger. A brain can be built by using these trust mechanism that can store neurons and recall the accurate information and trust the given experience of subjective and objective. The personal thinking chain of the user is connected with multiple factor authentication, qualified data that are self-common to humans are built safely. The data commons completely minimizes the human data siloes, also allows to keep the ownership for humans in privacy or to share the experience with no centralized authority and third party.

In future, this technology's augmented version can be used when more than two different humans experiencing the same moments, in a subjective perspective, this system can be used to reassemble the experiences of the users to become more objective during that time when the moment is happening. Thus, virtual simulation is created for the past memories, which can be used to see the memories from perspective of someone else. For this better understanding of mapping sensory experience and individual's emotions in the given memory is needed, this can even include senses (i.e., smell, sight, taste, and so on) in future BCT. So in the future, the sensory experience can be recorded with the help of wearable technologies. The brain's present state, biofeedback imaging, nerve implant's current status, or any sensor permitting multiple factor specific with fingerprints of specified human experience can be recorded. By utilizing all these technologies, leaning, rehabilitative protocols, recall, and decision-making can be improved.

The summary of applications that can adopt BCT in the health care/ biomedical to advance systems is given in Table 10.2.

10.4   CHALLENGES IN IMPLEMENTING BCT ON HEALTHCARE

The BCT is a new technology that is now being used in various industries [51]; it has high potential opportunities [21] and benefits [1]. But there are some challenges in this technology as shown in Figure 10.5; they are discussed in detail in the following sections.

10.4.1   PRIVACY AND SECURITY OF THE DATA

Data privacy and data security are the major crucial challenges in using BCT in the healthcare field [37]. The usage of BCT completely eliminates the concept of using third party to perform transactions [57,66]. There is no third party involved but the BCT allows all the users in that community to do verification process of the blockchain records, thus information becomes prone to risks in security and potential privacy [1]. There is a high privacy issue since every node in the blockchain community can get access to the data that is transmitted. In the authorization

---

TABLE 10.2

Summary of Applications in Blockchain Technology in Healthcare Industry

Application of Blockchain Technology	Summary
E-Health record system	A blockchain-based distributive ledger digital EHR system gives guarantee for integrity from generation of data process to data retrieval process with no human mediation.
Medical fraud deduction system	The blockchain consists of immutable feature that helps in deducting the fraud also does not allow any modification or duplication of transaction. This leads transaction to be secure and transparent.
Clinical research	Decentralized framework based on blockchain allows clinical research information collaboration for safe and secure sharing of the data with research groups.
Pharmaceutical industry research	Blockchain has the power to track detailed information on each and every step in the supply chain of pharmaceutical, such as medicine origination, compounds of medicine, and ownership that are deducted frequently in all the stages in order to avoid stealing or forgery of drugs.
Neuroscience	The innovation in blockchain leads to the development of many new upcoming applications such as brain thinking, brain stimulation, and brain augmentation. A huge medium is required to store the digitalized version of human brain, thus blockchain is used.

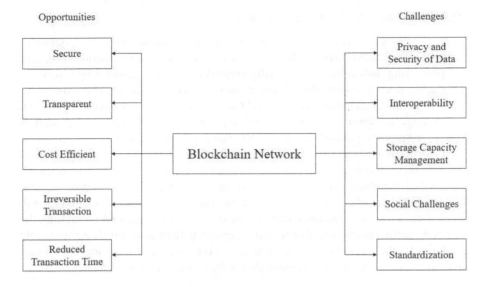

FIGURE 10.5 Opportunities and challenges in implementing blockchain technology on healthcare.

process, if the third party is absent, then the patient is responsible to choose more than one person to represent themselves who has the ability to access the data or clinical and medical records of the patients if there is an emergency situation occurs. In the BCT, the person who represents a patient has the ability to allow access for a set of users to data of the same patient; this process may lead to huge data security and data privacy threats. If the data and information of the patients are protected by mechanisms of high security, then it may result in difficulties to transfer data between different blocks, so receivers would be resulted in accessing incomplete or limited data. Fifty-one percent attack [32] is considered as a potential attack in the blockchain network. The attackers are set of miners who own higher than half percentage of total blocks available in the network of blockchain. If the miners hold more than half percentage of the blocks, then they could easily attain the network authority and thus, they can also have the ability on preventing any transaction to happen, this can be done by not giving required permission to them. There are five different crypto digital currencies that are considered as victims by coin desk [26], to this attack in recent past. Additionally, there might be high sensitive data in the records of a patient that are not suitable to be recorded in blockchain.

10.4.2 Issues in Interoperability

Interoperability is one of the issues in the blockchain used in the healthcare industry. It helps the blockchain from many different services and communication providers to have an appropriate and seamless talk between each other. When the data is shared effectively, this challenge generates hindrances [10].

10.4.3 STORAGE CAPACITY MANAGEMENT

Storage capacity management is one of the challenges that occurs in the blockchain used in the healthcare industry. In order to carry out the data transaction recording and processing, blockchain was initially created only with limited scope, so there is no heavy storage system. As the time passes, the technology spreads to too many different domains including healthcare; thus, the challenges of storage are revealed. A very huge amount of data is generated daily in the healthcare industry, such as X-rays, test reports, patient's records, MRI scans, patient's health history, and many other images regarding the medical treatment of the patient and all other medical data, all these details and data need a massive and huge space [6,47] to store the data in the blockchain which is available to each node present in the network chain. The blockchain applications mostly based on transaction of data; thus, the growth of the database is in rapid speed rate. As the database size keeps on increasing, the searching and accessing speed of records keeps on getting slow, this is not a suitable condition for the transaction, as speed is one of the essential factors for the transaction. So the solution for this challenge should be resilient [42] and scalable.

10.4.4 SOCIAL CHALLENGE

Blockchain is in the developing or evolvement stage, thus it also faces many social challenges including cultural shifts and other challenges excluding technical challenges. A technology that has a vast difference from traditional technology is very much hard to accept and adopt. In the current era of digitalization, the healthcare industry is also shifting slowly toward the digitalization but still there is very a long way in order to completely transfer the industry based on blockchain or digitalization of data since they are not yet validated in aspects of the medical and healthcare industries. Doctors are hard to convince in switching to modern technologies instead of using paperwork, and this process needs a lot of effort and time. In the current world, the adaptation of blockchain is low in healthcare and medical sector and thus the policies remain untrusted. As there are so many challenges and threats, the blockchain still cannot be used as a solution to every issue in the field of healthcare.

10.4.5 STANDARDIZATION

As blockchain is still in the early stage of development, the implementation of blockchain in the practical situations in the industry of healthcare and medicine faces challenges in standardization. To maintain a standard international authority, it requires many well-certified and authenticated standards. The information exchange format, nature of data, and size of data can be evaluated only if there are some predefined standards in the applications of blockchain. Defining standards will scrutinize the data that are shared and acts as safety precautionary measures.

To understand more, identify the strength, identify the weakness, opportunities, examine and find the threats that occur in blockchain network of healthcare and medical domain can be analyzed with the analysis of SWOT approach as shown in Figure 10.6.

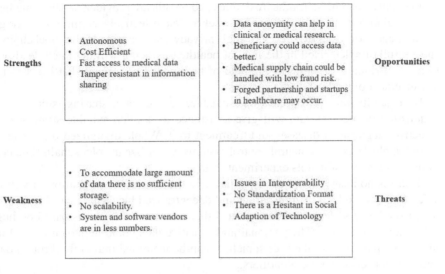

Strengths	<ul><li>Autonomous</li><li>Cost Efficient</li><li>Fast access to medical data</li><li>Tamper resistant in information sharing</li></ul>	<ul><li>Data anonymity can help in clinical or medical research.</li><li>Beneficiary could access data better.</li><li>Medical supply chain could be handled with low fraud risk.</li><li>Forged partnership and startups in healthcare may occur.</li></ul>	Opportunities
Weakness	<ul><li>To accommodate large amount of data there is no sufficient storage.</li><li>No scalability.</li><li>System and software vendors are in less numbers.</li></ul>	<ul><li>Issues in Interoperability</li><li>No Standardization Format</li><li>There is a Hesitant in Social Adaption of Technology</li></ul>	Threats

FIGURE 10.6   Blockchain in healthcare industry, SWOT analysis.

CONCLUSION AND FUTURE DIRECTIONS

The healthcare and medical industry sometimes waits behind various risks to receive the latest advancement because of the data integrity. In the near future, there might be a motivation to fill the gap of using blockchain in the field of medicine and healthcare. In the EEHR management, there can be some focal points. The usage of smart contracts in an application can take care of all the minor and major important exchanges of data within the healthcare and medical industry. Smart contracts are unbreakable chain consist of records or blocks, in which each block would get individual care with no breaking into healthcare framework. Healthcare systems can be converged with smart contracts of BCT that would help in resisting any duplicate copy production in centralized parent system. The research studies of blockchain should be provided with potential assistance in order to give many versions on scientific studies with time-stamped and verified details. The patients are allowed to control their own data with the help of smart contract. Similarly, the researchers are allowed to have sustained history on the findings of their researches; this can be done with the help of blockchain record documentation. BCT is compulsory in vast and huge industry such as pharmaceutical. All the report blocks spreading in blockchain in association with pharmaceutical can be deciphered. This can minimize the likelihood of harm or loss archive fabrications, guarantee straightforward and honest archive circulation, and expand the data preparation speed. Innovation itself has the capacity to control the interest points. The block which is already made cannot be erased and transformed, when using blockchain the data cannot be altered easily without noticing, fake or low-quality drugs can be easily rejected. These capabilities can be achieved by controlling each component in creation, logistics, and medicine sales blockchain.

Utilizing blockchain in healthcare domain practically will provide benefits to huge set of peoples like doctors, practitioners, research and development specialists, biomedical researchers, healthcare entities and healthcare providers to share medical or clinical knowledge, to share the huge set of data effectively, communicate or get recommendation from experts with high privacy and security. If the blockchain is successfully implemented in the field of healthcare, then it will definitely lead way to a lot of research avenues in the field of biomedical researches that leads to the advancement of medical field.

In a quality medical application, scalable, secure, safe, sharing, storage, and acquisition of medical data will give assistance to develop suitable strategies for effective diagnosis of diseases and treatment to it. Whole digitalized brain can be stored in blockchain, and neural control systems can utilize the blockchain. The field of neuroscience is still at its experimental stage.

There is no answer to a question, that how the data of a personal brain will be safe on the blockchain. A transparent and decentralized blockchain can prevent the data from being stolen or altered but still there are lots of concerns based on huge information collection. The personal and sensitive data could be stolen or sold for marketing purposes, so the user somehow can be identified indirectly through data patterns or pseudonymous identifiers.

There are lots of development in healthcare due to blockchain revolution such as telemedicine, similarly in the future, the medical field will be taken to the next level of development by minimizing configuration, monitoring cost, and maintaining a single central server and administration to manage medical data. Using blockchain in clinical works leads to process time minimization, because once the patient registers, complete information on the patient is collected and can be accessed anytime since distributed ledger is used. Also doctors get access to the patient's medical history in real time since the doctor has access to view patient's quality, authentic and original data so that no medical error can occur due to misunderstanding or miscommunication between the doctor and the patient. Also the data in the blockchain is transparent and the patient can easily get advice from another expert. The blockchain that consists of patient data and records connects lots of people around the world with the same medical conditions which will result in acceptance of patient feelings, improves health, gives support, willpower, and strength in order to fight with the disease. Also the patients have complete access to their data, and they have the power to decide to whom the data can be shared. The blockchain in healthcare sector will definitely maximize the improvement in the quality of the lifestyle and also make individuals engage in their own medical or healthcare activities.

REFERENCES

1. Alhadhrami, Z., Alghfeli, S., Alghfeli, M., Abedlla, J. A. and Shuaib, K. Introducing blockchains for healthcare. In Proceedings of the 2017 IEEE International Conference on Electrical and Computing Technologies and Applications (ICECTA), Ras Al Khaimah, UAE. doi: 10.1109/ICECTA.2017.8252043, 2017.
2. Angraal, S., Krumholz, H. M. and Schulz, W. L. Blockchain technology: Applications in health care. *Circ. Cardiovasc. Qual. Outcomes.* 10(9), e003800. doi: 10.1161/CIRCOUTCOMES.117.003800, 2017.

3. Aranki, D., Kurillo, G., Mani, A., Azar, P., Gaalen, J. V., et al. A telemonitoring framework for android devices. Connected Health: Applications, Systems and Engineering Technologies (CHASE), 2016 IEEE First International Conference on. IEEE, 2016.

4. Azaria, A., Ekblaw, A., Vieira, T. and Lippman, A. MedRec: Using blockchain for medical data access and permission management. In Proceedings of the 2nd International Conference on Open and Big Data (OBD 16), Vienna, Austria. 25–30. doi: 10.1109/OBD.2016.11, 2016.

5. Benchoufi, M., Porcher, R. and Ravaud, P. Blockchain protocols in clinical trials: Transparency and traceability of consent. *F1000Research*. 6, 66. doi: 10.12688/f1000research.10531.5, 2017.

6. Bennett, B. Blockchain HIE overview: A framework for healthcare interoperability. *Telehealth Med. Today*. 1–6. Available at: https://pdfs.semanticscholar.org/1756/5afb7c f33f49985564a46dfcd97089b4bc9f.pdf, 2017.

7. Berman, M. and Fenaughty, A. Technology and managed care: Patient benefits of tclc-medicine in a rural health care network. *Health Econ*. 14(6), 559–573. doi: 10.1002/hec.952, 2005.

8. Bhatti, A., Siyal, A. A., Mehdi, A., Shah, H., Kumar, H. and Bohyo, M.A. Development of cost-effective tele-monitoring system for remote area patients. In Proceedings of the 2018 IEEE International Conference on Engineering and Emerging Technologies (ICEET), Lahore, Pakistan. 22–23, 2018.

9. Bockstael, N. E. The use of random utility in modeling rural health care demand: Discussion. *Am. J. Agri. Econ*. 81, 692–695, 1999.

10. Boulos, M. N., Wilson, J. T. and Clauson, K. A. Geospatial blockchain: Promises, challenges, and scenarios in health and healthcare. *Int. J. Health Geogr*. 25(2018). Available at: https://ij-healthgeographics.biomedcentral.com/articles/10.1186/s12942-018-0144-x, 2018.

11. Brandon, R. M., Podhorzer, M. and Pollak, T. H. Premiums without benefits: Waste and inefficiency in the commercial health insurance industry. *Int. J. Health Serv*. 21(2), 265–283. doi:10.2190/H824-R263-YL47-WRQD, 1991.

12. Buterin, V. A Next-Generation Smart Contract and Decentralized Application Platform; White Paper; Ethereum Foundation (Stiftung Ethereum): Zug, Switzerland. Available at: http://blockchainlab.com/pdf/Ethereum_white_paper-a_next_generation_smart_contractand_decentralized_application_platform-vitalik-buterin.pdf, 2014.

13. Castaneda, C., Nalley, K., Mannion, C., Bhattacharyya, P., Blake, P., et al. Clinical decision support systems for improving diagnostic accuracy and achieving precision medicine. *J. Clin. Bioinform*. 5, 4. doi: 10.1186/s13336-015-0019-3, 2015.

14. Chen, Y., Ding, S., Xu, Z., Zheng, H. and Yang, S. Blockchain-based medical records secure storage and medical service framework. *J. Med. Syst*. 43(1), 5. doi:10.1007/s10916-018-1121-4, 2018.

15. Cyran, M. Blockchain as a Foundation for Sharing Healthcare Data. *Blockchain in Healthcare Today*, 1. https://doi.org/10.30953/bhty.v1.13, 2018.

16. Desalvo, K. and Galvez, E. *Connecting Health and Care for the Nation: A Shared Nationwide Interoperability Roadmap*;Washington, DC: Office of the National Coordinator for Health Information Technology, 2015.

17. Downing, N. L., Adler-Milstein, J., Palma, J. P., Lane, S., Eisenberg, M., et al. Health information exchange policies of 11 diverse health systems and the associated impact on volume of exchange. *J. Am. Med. Inform. Assoc*. 24(1), 113–122. doi: 10.1093/jamia/ocw063, 2017.

18. Dubovitskaya, A., Xu, Z., Ryu, S., Schumacher, M. and Wang, F. Secure and trustable electronic medical records sharing using blockchain. In AMIA Annual Symposium Proceedings; American Medical Informatics Association: Washington, DC, USA. Vol. 2017, 650–659, 2017.

19. Ekblaw, A., Azaria, A., Halamka, J. D and Lippman, A. A. Case study for block-chain in healthcare: "MedRec" prototype for electronic health records and medical research data. In Proceedings of the IEEE Open and Big Data Conference, Vienna, Austria. Vol. 13, 1–14. Available at: http://www.sbrc2018.ufscar.br/wp-content/uploads/2018/04/07-181717-1.pdf, 2016.

20. Emerging technology from the arXiv: Google's secretive DeepMind startup unveils a neural Turing machine. M.I.T. Technology Rev. Available at: http://www.technologyreview.com/view/532156/googlessecretive-deepmind-startup-unveils-a-neural-turing-machine/, accessed July 22, 2015.

21. Fernández-Caramés, T. M. and Fraga-Lamas, P. A. Review on the use of blockchain for the internet of things. *IEEE Access* 6, 32979–33001, 2018. Available at: https://ieeexplore.ieee.org/stamp/stamp.jsp?arnumber=8370027, 2018.

22. Gomez, E. J., delPozo, F. and Hernando, M. E. Telemedicine for diabetes care: The DIABTel approach towards diabetes telecare. *Med. Inform. (Lond)*. 21, 283–295, 1996.

23. Gorman, L. The history of health care costs and health insurance. *Wisconsin Policy Res. Inst. Rep.* 19(10), 1–31. Available at: http://www.wpri.org/BI-Files/Special-Reports/Reports-Documents/Vol19no10.pdf, 2006.

24. Griebel, L., Prokosch, H. U., Köpcke, F., Toddenroth, D., Christoph, J., et al. A Scoping review of cloud computing in healthcare. *BMC Med. Inform. Decis. Mak.* 15, 17. doi: 10.1186/s12911-015-0145-7, 2015.

25. Griggs, K. N., Ossipova, O., Kohlios, C. P., Baccarini, A. N., Howson, E. A. and Hayajneh, T. Healthcare blockchain system using smart contracts for secure automated remote patient monitoring. *J.Med. Syst.* 42(7), 130. doi: 10.1007/s10916-018-0982-x, 2008.

26. Hertig, A. Blockchain's Once-Feared 51 Percent Attack Is Now Becoming Regular. Available at: https://www.coindesk.com/blockchains-feared-51-attack-now-becoming-regular/, 2018.

27. Heston, T. A case study in blockchain health care innovation. *Int. J. Curr. Res.* 9(11), 60587–60588. doi: 10.22541/au.151060471.10755953, 2017.

28. Horn, G., Martin, K. M. and Mitchell, C. J. Authentication protocols for mobile network environment value-added services. *IEEE Trans. Vehicular Tech.*, 51(2), 383–392, 2002.

29. Houston, M. S., Myers, J. D., Levens, S. P., McEvoy, M. T., Smith, S. A., et al. Clinical consultations using store-and-forward telemedicine technology. In *Mayo Clinic Proceedings*; Rochester, MN: Elsevier. Volume 74, 1999.

30. How Blockchain Will Revolutionise Clinical Trials. Available at: https://pharmaphorum.com/views-and-analysis/how-blockchain-will-revolutionise-clinical-trials-clinical-trials/, 2018.

31. Ianculescu, M., Stanciu, A., Bica, O. and Neagu, G. Innovative, adapted online services that can support the active, healthy and independent living of ageing people. A case study. *Int. J. Econ. Manag. Syst.* 2, 321–329, 2017.

32. Investopedia "Blockchains". Available at: https://www.investopedia.com/terms/1/51-attack.asp, 2019.

33. Ivan, D. Moving toward a blockchain-based method for the secure storage of patient records. In ONC/NIST Use of Blockchain for Healthcare and Research Workshop; ONC/NIST: Gaithersburg, MD, USA. Available at: https://www.healthit.gov/sites/default/files/9-16-drew_ivan_20160804_blockchain_for_healthcare_final.pdf, 2016.

34. Jiang, S., Cao, J., Wu, H., Yang, Y., Ma, M. and He, J. Blochie: A blockchain-based platform for healthcare information exchange. In Proceedings of the 2018 IEEE International Conference on Smart Computing (SMARTCOMP), Taormina, Italy. 18–20, 2018.

35. Kevin, A. C., Breeden, E. A., Davidson, C. and Mackey, T. K. Leveraging Blockchain Technology to Enhance Supply Chain Management in Healthcare: An exploration of challenges and opportunities in the health supply chain. Blockchain Healthc. *Today*. Available at:https://doi.org/10.30953/bhty.v1.20, 2018.
36. Kuo, M. H., Kushniruk, A. and Borycki, E. Can cloud computing benefit health services?: ASWOT analysis. *Stud Health Technol Inform*. 169, 379–83, 2011.
37. Kuo, T. T., Hsu, C. N. and Ohno-Machado, L. ModelChain: Decentralized privacy-preserving healthcare predictive modeling framework on private blockchain networks. arXiv 2016, arXiv:1802.01746, 2016.
38. Kuo, T. T., Kim, H. and Ohno-Machado, L. Blockchain distributed ledger technologies for biomedical and health care applications. *J. Am. Med. Inform. Assoc*. 24(6), 1211–1220. doi: 10.1093/jamia/ocx068, 2017.
39. Mandl, K. D., Markwell, D., MacDonald, R., Szolovits, P. and Kohane, I. S. Public standards and patients' control: How to keep electronic medical records accessible but private. *BMJ*. 322(7281), 283–287. doi:10.1136/bmj.322.7281.283, 2001.
40. Markov, A. Use of Blockchain in Pharmaceutics and Medicine. Available at: https://miningbitcoinguide.com/technology/blokchejn-v-meditsine/
41. Mauri, R. Blockchain for Fraud Prevention: Industry Use Cases. Available at: https://www.ibm.com/blogs/blockchain/2017/07/blockchain-for-fraud-prevention-industry-use-cases/, 2017.
42. McKinlay, J. Blockchain: Background Challenges and Legal Issues; DLA Piper Publications: London, UK, Available at: https://www.dlapiper.com/en/denmark/insights/publications/2017/06/blockchain-background-challenges-legal-issues/, 2016.
43. Mettler, M. M. Blockchain technology in healthcare: The revolution starts here. In Proceedings of the 2016 IEEE 18th International Conference on e-Health Networking, Applications and Services (Healthcom), Munich, Germany. doi: 10.1109/HealthCom.2016.7749510, 2016.
44. Moe Alsumidaie. Blockchain Concepts Emerge in Clinical Trials, Applied Clinical Trials. Available online: http://www.appliedclinicaltrialsonline.com/blockchain-concepts-emerge-clinical-trials, 2018.
45. Nichol, P. B. Blockchain Applications for Healthcare. Available at: http://www.cio.com/article/3042603/innovation/blockchain-applications-for-healthcare.html, 2016.
46. Nugent, T., Upton, D. and Cimpoesu, M. Improving data transparency in clinical trials using blockchain smart contracts. *F1000Research*. 5, 2541. doi: 10.12688/f1000research.9756.1, 2016.
47. Pirtle, C. and Ehrenfeld, J. Blockchain for healthcare: The next generation of medical records?*J. Med. Syst*. 42(9), 172. doi: 10.1007/s10916-018-1025-3, 2018.
48. Plotnikov, V. and Kuznetsova, V. Theprospects for the use of digital technology "blockchain" in the pharmaceutical market. In MATEC Web of Conferences; EDP Sciences: London, UK. Vol 193. doi: 10.1051/matecconf/201819302029, 2018.
49. Ross, J. S. The committee on the costs of medical care and the history of health insurance in the united states. *Einstein Quarterly*. 19, 130, 2002.
50. Rudin, R. S., Motala, A., Goldzweig, C. L. and Shekelle, P. G. Usage and effect of health information exchange: A systematic review. *Ann Intern Med*. 161, 803–811, 2014.
51. Shae, Z. and Tsai, J. J. On the design of a blockchain platform for clinical trial and precision medicine. In Proceedings of the 2017 IEEE 37th International Conference on Distributed Computing Systems (ICDCS), Atlanta, GA, USA. Vol. 1, 1972–1980, 2017.
52. Shubbar, S. Ultrasound Medical Imaging Systems Using Telemedicine and Blockchain for Remote Monitoring of Responses to Neoadjuvant Chemotherapy in Women's Breast Cancer: Concept and Implementation. Master's Thesis, Kent State University, Kent, OH, USA, 2017.

53. Swan, M. Blockchain thinking: The brain as a decentralized autonomous corporation. *IEEETechnol. Soc.Mag.* 34(4), 41–52. doi: 10.1109/MTS.2015.2494358, 2015.
54. Sylim, P., Liu, F., Marcelo, A. and Fontelo, P. Blockchain technology for detecting falsified and substandard drugs in distribution: Pharmaceutical supply chain intervention. *JMIR Res. Protoc.* 7(9), e10163. doi: 10.2196/10163, 2018.
55. Taylor, P. Applying Blockchain Technology to Medicine Traceability. Available at: https://www.securingindustry.com/pharmaceuticals/applying-blockchain-technology-to-medicinetraceability/s40/a2766/, 2016.
56. Trujllo, G. and Guillermo, C. The Role of Blockchain in the Pharmaceutical Industry Supply Chain as a Tool for Reducing the Flow of Counterfeit Drugs. Ph.D. Thesis, Dublin Business School, Dublin, Ireland, 2018.
57. Tschorsch, F. and Scheuermann, B. Bitcoin and beyond: A technical survey on decentralized digital currencies. *IEEE Commun. Surveys Tutorials.* 18(3), 2084–2123, 2016.
58. Wang, F. Y. and Wong, P. K. Intelligent systems and technology for integrative and predictive medicine: An ACP approach. *ACM Trans. Intell. Syst. Technol.* 4(2), 1–16, 2013.
59. Wang, S., Wang, J., Wang, X., Qiu, T., Yuan, Y., et al. Blockchain-powered parallel healthcare systems based on the ACP approach. *IEEE Trans. Comput. Soc. Syst.* 99, 1–9. doi: 10.1109/TCSS.2018.2865526, 2018.
60. Witchey, N. J. Healthcare transaction validation via blockchain proof-of work, systems and methods. US patent US20150332283 A1, 2015.
61. Wood, G. Ethereum: A Secure Decentralised Generalised Transaction Ledger; Yellow Paper; Ethereum Foundation (Stiftung Ethereum): Zug, Switzerland. Available at:https://gavwood.com/paper.pdf, 2014.
62. Xia, Q., Sifah, E. B., Asamoah, K. O., Gao, J., Du, X. and Guizani, M. MeDShare: Trust-less medical data sharing among cloud service providers via blockchain. *IEEE Access* 5, 14757–14767. doi: 10.1109/ACCESS.2017.2730843, 2017.
63. Yue, X., Wang, H., Jin, D., Li, M. and Jiang, W. Healthcare data gateways: Found healthcare intelligence on blockchain with novel privacy risk control. *J. Med. Syst.* 40(10), 218. doi: 10.1007/s10916-016-0574-6, 2016.
64. Zhang, J., Xue, N. and Huang, X. A Secure System for Pervasive Social Network Based Healthcare. *IEEE Access.* 4, 9239–9250, 2016.
65. Zhang, P., White, J., Schmidt, D. C., Lenz, G. and Rosenbloom, S. T. Fhirchain: Applying blockchain to securely and scalably share clinical data. *Comput. Struct. Biotechnol. J.* 267–278. doi: 10.1016/j.csbj.2018.07.004, 2018.
66. Zheng, Z., Xie, S., Dai, H., Chen, X. and Wang, H. An overview of blockchain technology: Architecture, consensus, and future trends. In Proceedings of the 2017 IEEE International Congress on Big Data (BigDataCongress), Honolulu, HI, USA. Available at: https://www.semanticscholar.org/paper/An-Overview-of-Blockchain-Technology%3A-Architecture%2C-Zheng-Xie/ee177faa39b981d6dd21994ac33269f3298e3f68, 2017.

11 Blockchain Theories and Its Applications

Jaipal Dhobale and Vaibhav Mishra
ICFAI Business School

CONTENTS

11.1 INTRODUCTION

Blockchain! Now a days, it has become a buzzword. What is special about it? Why and how blockchain is a breakthrough technology? Before finding answers to these questions, we will understand the different problems faced by the enterprises during the time. Prior to 2008, enterprises specifically working with database applications over the networks were facing problems like nonrepudiation, maintaining transparency, unauthorized updation/alteration, and security of these transactions against unauthorized utilization. These enterprises were heavily dependent on third parties to tackle these issues. Although we are discussing about enterprises in general, specifically financial institutions were the major concern about these problems and their solutions. This is the reason in 2008, blockchain technology was introduced to the world in the form of cryptocurrency application [1].

Blockchain is a technology that breaks these process hurdles and gets through the transactions successfully in a cost-effective and efficient way. This is the reason why it is a breakthrough technology. It is a chain of uneditable blocks of fixed sizes. These blocks are holding the transaction data up to 1MB [2]. Before preparing the blocks of transactions, these transactions were approved by the users in the network. Such prepared blocks are then attached to the previously created blocks to continue as a chain of blocks. Transaction information after adding every block in the chain is distributed across networks and is available to everyone. This process is carried out without intermediaries like financial institutions, government and non-government organizations. Blockchain carries all the work with encryption-based hash values. All the users in the network are sharing the details on the basis of peer-to-peer (P2P) network, where all the nodes are working at the same level. Every user in the blockchain network saves local copy of the blockchain. When consensus validates the transaction in the group, then it will be added as the block of transaction in the blockchain. All this is done by taking care of privacy of the resources involved in the transaction. This technology involves "smart contracts" [3] automatically through computer programs.

The specialty of the blockchain is the availability of this uneditable information to everyone giving transparency to the transactions across business applications. This is really beneficial to both enterprises and their stakeholders, where enterprises are delivering values through networks. It is like Distributed Ledger Technology (DLT) [4] but a bit different from it [5]. Both these technologies are working with decentralized networks, but in DLT, the owner/user is having more control over the networks, and information in the network is stored on different servers. In DLT, every node in the network will have its ledger and it will get updated immediately after the transaction at node level independently. In blockchain after approving transaction by consensus, blocks are prepared in cryptic format, and these are added to the chains of blocks based on certain hash functions to form the blockchain. We can say blockchain is one of the DLT. Figure 11.1 depicts the philosophy.

11.2 BLOCKCHAIN WORKING

Blockchain is decentralized and distributed ledger of blocks. These blocks are nothing but validated transactions by the participating nodes in the networks that cannot be modified or deleted after creating blocks. Blocks in the blockchain hold:

- Transaction data – Block version number, timestamp, and transaction details.
- Signature of the last block.
- A Nonce is the abbreviation of a number used only once and is a random string of number generated by miner, which is appended to encrypted block.

Figure 11.2 depicts the typical block in the blockchain.

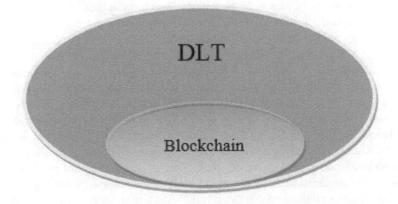

FIGURE 11.1   DLT and blockchain relationship.

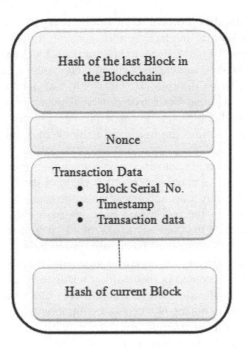

FIGURE 11.2 Typical block structure in blockchain.

Sufficient miners from the network nodes approve these transactions. Miner node with sufficient computational power can find the eligible signature to generate the blocks. These miner nodes use special hardware to attain computational power. This process is nothing but mining process. Mining process is done with proof-of-work (PoW) philosophy/protocol [3]. PoW is a protocol that calculates random hash value data from the block, which is very complicated and time-consuming computation process. After approval of transactions from miner, block will be appended to the chain as a new block. Miners upon creating new blocks will be rewarded with some sort of incentives in the form of Bitcoin.

As PoW is complicated computation and consumes a lot of computational resources, "Proof of Stake" (PoS) philosophy/protocol [6] was used as an alternate option, where miners were rewarded based on how much stake they are holding while mining the block.

"Delegate Proof of Stake" (DPoS) [7] is another form of PoS [8] in which blockchain users elect their delegate. This elected delegate node can generate and validate the blocks.

"Practical Byzantine Fault Tolerance" (PBFT) [7] is another protocol to mine the block. This protocol is used to take care of malicious activity in the blockchain. Blocks are mined in the rounds. In each round, rule-based new primary will be selected. This primary is responsible for mining the block. This process will be conducted in three steps – pre-prepared, prepared, and commit. To prevent malicious activity during each step in a round, primary has to receive consent of two-thirds of total nodes.

"Ripple" [7] is another protocol to generate nodes. It will create trusted sub-network in the blockchain. and based on the consent of these trusted subnetwork nodes, it will create the block. Under this protocol, blockchain network will be divided into two parts – servers and clients. The role of server is to carryout consensus process, while the role of clients is to transfer funds. Each server in the chain maintains Unique Node List (UNL). The server asks UNL about the transaction validity, if it 80% and above, UNL agrees and then transaction will be validated and added into the blockchain. Blockchain process is depicted in Figure 11.3.

Blocks may be generated simultaneously as multiple miners may find a suitable nonce at the same time. It may create another chain in the blockchain but defined protocol will consider only longer chains as authentic chain.

Each added block in the blockchain contains hash value from the previous block. If we try to alter the contents of the block, it will change hash value of the current block. This change of hash value mismatches with the hash value reflected in the immediate block making it unchain from the blockchain. If this altered transaction is accepted in the chain, it needs to be chained in subsequent blocks by changing hash value of every succeeding block in the chain, which is highly impossible. This makes blockchain as immutable [4]. In this way, block-chain first validates the transaction and safeguards it, and at last, these entries are preserved as records. Transaction in the blockchain is protected using digital signature based on public and private key mechanism. Blockchain structure is depicted in Figure 11.4.

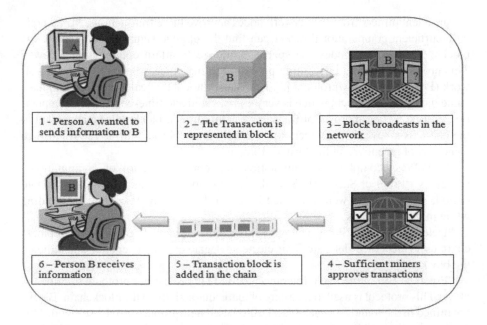

FIGURE 11.3   Adding block in the blockchain.

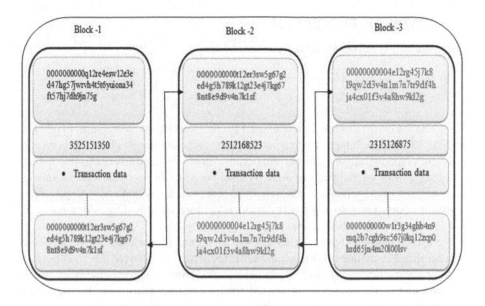

FIGURE 11.4   Blockchain structure.

Blockchain participants and their roles [9]:

1. Regulating Authority – Blockchain user who is having special permission to regulate blockchain transactions. These users will deal with the block-chain regulations as per the decided policy. These users are restricted from carrying out blockchain transactions.
2. Blockchain Developer – These are the individuals who will develop various applications and protocols to carry out blockchain transactions. They are the hardcore programmers, and with their programming skills, they provide a suitable platform to blockchain users to operate. These are the users who will take care of the data sources and blockchain platforms with added skill sets.
3. Blockchain Network Administrators – These are the individuals with network administration skills. They are the one who will manage blockchain network administration by providing user accessibility to the blockchain resources as per defined policy.
4. Blockchain Users – They are the one who carries blockchain transactions. These individuals avail facilities offered by the blockchain architecture to carryout blockchain transactions. These users are unaware of the technicalities on offer to carry blockchain transactions.
5. Certificate Authorities – As per policy, these individuals will pervade the transaction certificates to the users.

11.3 TYPES OF BLOCKCHAIN

There are two types of blockchain platforms [3]:

1. Permissioned blockchain.
2. Permissionless blockchain.

This is based on the consensus model used for the working of blockchain. In permissioned [10–12] blockchain, participants need consent to join the chain. This type of platform is specifically used by private organization with centrally controlled nature. Governing of blockchain can be done by a single entity or group of representatives. Transactions of permissioned blockchain platforms are available only to permitted representatives. The role of the participants is decided by providing additional security layer in the blockchain architecture.

Permissionless platform is also known as public blockchain, which is purely decentralized in nature also known as public ledgers. It does not require any permission for the participants to join blockchain. Participants can download open-source protocol and be the part of blockchain. Permissionless platforms are having tokens to provide incentives to the participants. Based on the application usages, it can also be divided into three types [12,13]:

1. Public blockchain.
2. Private blockchain.
3. Consortium blockchain.

Public blockchain is nothing but permissionless blockchain where there is no restriction on participants and validators. Private blockchain is permissioned blockchain where only pre-approved participants are permitted to participate and/or validate the transaction. Consortium blockchain is also known as hybrid blockchain because the information is being shared like public blockchain but it can restrict the nodes that can access the blockchain.

11.3.1 CHARACTERISTICS OF BLOCKCHAIN [12]

Based on the above discussion, we enlist the characteristics of blockchain:

1. Transparency in maintaining data in the form of transaction.
2. Immutability – transaction once recorded is time stamped and is highly impossible to alter.
3. Record persistency – transactions are checked and validated by the network group members.
4. Auditability – information in the blocks are easier to trace and verify because of chain structure with time stamping.
5. Decentralization control – distributed ledger mechanism.
6. Low transaction cost – removal of intermediary's role subsequently reduces transaction costs.

7. Real-time data capturing and updation.
8. Security using encryption technique.

11.3.2 CHALLENGES OF BLOCKCHAIN [14–16]

There are certain challenges that we need to deal with while implementing the blockchain. Those are:

1. 51% attack – This is the kind of attack in which blockchain users form group as "mining pool" to mine more blocks. This mining pool holds maximum computing power. If this holding power is more than 51% computational capacity of the blockchain, then it will take control of the blockchain. After getting control by the group, they can modify transactions, can stop miners from mining blocks, can stop transaction validating process. These types of attacks were more feasible in the past when network size was smaller.
2. Fork problem – This problem occurs when blockchain was upgraded from old version to new version. After upgradation, blockchain nodes can be divided into new nodes and old nodes. This problem occurs when non-upgraded nodes will not validate upgraded nodes or non-upgraded nodes will not follow the new consensus regulations. Because of this divergence took place in the blockchain nodes, if it is permanent, it is called as hard fork, and if it is temporary, it is called soft fork.
3. Scalability – Blockchain network keeps on growing day by day. With an increase in the blockchain, data and resources keep on adding burden on the system. Because of this, nodes will take more time to synchronize data and carry out complex computation. It will subsequently affect the effective working of the blockchain system.

11.4 HISTORY AND DEVELOPMENT OF BLOCKCHAIN

Stuart Haber and W. Scott Stornetta introduced blockchain in 1991 but real utilization of this technology came in view during the year 2008 [17] in the form of cryptocurrency application by unknown entity as an individual or small group called Santoshi Nakamoto. Research paper "Bitcoin: A Peer-To-Peer Electronic Cash System" by Santoshi Nakamoto explained the application of blockchain to transfer Bitcoin between network group members without using third-party involvement like financial institutions.

After the introduction of this new mechanism, the first block of the blockchain was released. It is called Block 0, which is genesis block of blockchain and is a special case block, which does not contain address of previous block in blockchain technology. This is basic building block of the blockchain technology. Genesis block is developed by the originators of the blockchain. From records, it is reflected that it took six days to mine genesis block while average timestamp for mining a block in the blockchain is 10 minutes. Genesis block consists of 50 Bitcoin rewards. It is not clear whether it was kept intentionally or unintentionally. Using this open-source protocol, anyone across the world can participate in the blockchain network by entering into the contract.

11.4.1 Applications of Blockchain

Since its inception, the financial industry is the major beneficiary of blockchain technology; the most popular application is cryptocurrency. Looking after its potential, other industries are also exploring opportunities. We are discussing herewith financial services, government uses, health care, industrial applications, and Internet of Things (IoT) only.

1. Financial services [18,19] – Blockchain is effective in taking care of the financial services like commercial financing, trade finance, cross-board transactions, and digital payments. It can play important role in providing sustainable economic growth. To provide end-to-end visibility of financial resources, blockchain is proven technology.

   Organizations can trace and maintain their commercial financing with suppliers and its credit details effectively. It will help in reducing number of disputes and time to resolve disputes. Businesses need to interact with various legal entities and fulfill legal requirements. Blockchain will help businesses in the area to maintain contracts and keep updating the same with various business entities. It will simplify the business processes.

   MNCs need access to multiple currencies along with facility to maintain transactions in the local currencies to run businesses. Such transactions and its reconciliations will be taken care effectively with the help of blockchain technology. It will provide visibility of transactions and balances across multiple locations.

   Banking institutions are using blockchain as a disruptive technology to resolve the traditional banking problems such as nonrepudiation, transparency, traceability in a cost-effective way. R3 [18], like blockchain application, is used in the banks to settle cryptocurrency payments, general banking applications, loan management, and financial auditing. Crowdfunding [20] is another business functions that can be thought of through blockchain technology. Various blockchain applications such as Prediction Marketplace systems, BitShares 2.0, Plasma, and Global Payments Steering Groups are already in use to deal with banking services.

2. Government Uses [18] – Government provides services to the citizens and enterprises. Different stakeholders such as beneficiaries, bureaucrats, policy makers, and community organizations need frequent interactions and exchange of information. Government needs to maintain records of these transactions as proof of ownership and proof of existence. Blockchain could provide better solution to the existing procedures. Blockchain will enhance accountability and safety of the records. It will enhance operations performance by automating governance. Services like citizen identification certificates, maintaining the records of the public assets, tracking its uses, record keeping and maintaining legal documents related to the public and private entities, various permissions issuing to the stakeholders, tax payments can be taken care by blockchain applications in a better way. Blockchain could help bitterly in integration of physical assets and its integration with digital contents. This will

help in reducing bureaucracy and improving the operational efficiency of the government services. Online voting could be possible along with the integration of government services as blockchain records are immutable, traceable, decentralized public ledgers along with great transparency.

3. Healthcare [3] – Healthcare is another area where blockchain would be effective because of its features such as immutability, distributed public ledger, transparency, availability of required information in a secure way by maintaining privacy. Blockchain will sort the problem of stakeholder's integration in healthcare supply chain management effectively. Healthcare functions such as maintaining patient healthcare records, clinical trial, public health scheme records, drug counterfeiting, automating health insurance mechanism, and accounting would be taken care by blockchain technology in a better way. While performing these operations, privacy and security where at most concern, which blockchain will take care of?

4. Industrial applications [18] –Business or industry tracking of movement of goods under process or processed goods is very important in order to attain the business objectives. Similarly, tracking the various business services and its optimization is very important to satisfy stakeholders. Business operations such as purchase management, customer relationship management, supply chain management, production and operations management need to be taken care. Visibility of the production process and its associated services could be well taken care with the help of blockchain technology. Blockchain technology features make these operations effective and efficient. This will provide various opportunities in the area discussed above and support business to flourish.

5. IoT applications [18] – This is one of the most challenging but yet growing field in which smart devices interact with each other using the Internet. These devices generate huge data. The data that is generated by these devices in the last two to three years is almost equal to the data that is already the world is having. Major problem of this field is security of data generated by different smart devices within distributed nature of wireless networks. As we have discussed that blockchain is distributed public ledger, it would best take care of this huge data and its security. Blockchain will also help in payment settlement mechanisms among these smart devices throughout the world in cost-effective and more secure way.

These are the potential areas where the opportunities of blockchain can be explored. There are certain challenges of the blockchain technology itself along with area-specific challenges which we need to address.

REFERENCES

1. Nakamoto Santosh, "Bitcoin: A Peer-to-Peer Electronic Cash System," December 17, 2019 18.22, https://www.bitcoin.org.
2. Baliqa Arati, "Understanding Blockchain Consensus Models," *Persistent* Systems *Ltd*, (2017):3–14

3. BVS Girish, "Blockchain Technology: Concepts," *Sasken Technologies Ltd*, December 20, 2019, 18.00, https://www.sasken.com/insights/whitepaper.

4. Cohen James, "The blockchain revolution: The ultimate financial services industry disrupter," *Hewlett Packard Enterprise*, 2016, 20–12–2019, 11.08.

5. Belin Oliver, "The Difference between Blockchain and Distributed Ledger Technology," December 18, 2019, 15.22, https://tradeix.com/distributed-ledger-technology.

6. Capital Hexayurt, "Building the Hyperconnected Future on Blockchains," *World Government Summit, February 2017*, December 17, 2019, 19.02, https://www.world-governmentsummit.org/observer/reports/2017/detail/building-the-hyperconnected-future-on-blockchains.

7. Zheng Zibin, Shaoan Xie, Hong-Ning Dai, Xiangping Chen, Huaimin Wang, "Blockchain challenges and opportunities: A survey", *International Journal of Web and Grid Services*, Vol 14, No. 4 (2018): 353–375.

8. Casinoa Fran, Thomas K. Dasaklisb, Constantinos Patsakisa, "A systematic literature review of blockchain-based applications: Current status, classification and open issues," *Telematics and Informatics*, 36 (2019):55–81.

9. Gupta Manav, *Blockchain for dummies,* Hoboken, NJ: John Wiley & Sons, Inc., 2017.

10. Contri Bob, Rob Galaski, "Over the horizon—Blockchain and the future of financial infrastructure-," *Deloitte*, December 17, 2019, 14.5, https://www2.deloitte.com/au/en/pages/financial-services/articles/over-horizon-blockchain-future-financial-infrastructure.html.

11. Michael Croby, Nachiappan, Pradan Pattanayak, Sanjeev Verma and Vignesh Kalyanaraman, "Blockchain technology: Beyond bitcoin," *Applied Innovation Review*, 2 (2016): 6–19.

12. Ganne Emmanuelle, "Can Blockchain revolutionize international trade?" *World Trade Organization*, Geneva 2, December 17, 2019, 12.28, https://www.wto.org.

13. Houben Robby, Alexander Snyers, "Cryptocurrencies and blockchain—Legal context and implications for financial crime, money laundering and tax evasion", European Parliament, 2018.

14. Kassab Mohamad, Jonna DeFranco, Tarek Malas, Valdemar Vicente Graciano Netoand Giuseppe Destefanis, "Blockchain: A Panacea for Electronic Health Records?" December 17, 2019, 18.15. https:/:www.researchgate.net/publication/331729440.

15. Ko Vanessa, Andrej Verity (OCHA), "Blockchain for the humanitarian sector: Future opportunities," *DH Network*, 2016. https://blog.veritythink.com/post/153969586544/blockchain-for-the-humanitarian-sector-future.

16. Koteska Bojana, Elena Karafiloski, Anastas Mishev, "Blockchain Implementation Quality Challenges: A Literature Review," (Full Paper, SQAMIA 2017: *6th Workshop of Software Quality, Analysis, Monitoring, Improvement and Applications, Belgrade, Serbia*, 11–September 13, 2019)

17. White Mark, Jason Killmeyer, Bruce Chew, "Will Blockchain Transform the Public Sector? Blockchain Basics for Government," A report from the Deloitte center for government insights, December 17, 2017, 14.53. https://www2.deloitte.com/us/en/insights/industry/public-sector/understanding-basics-of-blockchain-in-government.html.

18. Nofer Michael, Peter Gomber, Oliver Hinz, Dirk Schiereck, "Blockchain," *Business & Information Systems Engineering*, 59 (3), (2017): 183–187.

19. Stagnaro Chet, "Innovative Blockchain Uses in Health Care," *Freed Associates*, December 17, 2019, 15.36, https://www.freedassociates.com/wp-content/uploads/2017/08/Blockchain_White_Paper.pdf".

20. Sompolinsky Yonatan, Aviv Zohar, "Accelerating Bitcoin's Transation Processing Fast Money Grows on Trees, Not Chains," December 17, 2019, 18.58, https://pdfs.semanticscholar.org/4016/80ef12c04c247c50737b9114c169c660aab9.pdf.

12 Building Permissioned Blockchain Networks Using Hyperledger Fabric

K. Varaprasada Rao, Mutyala Sree Teja,
P. Praneeth Reddy, and S. Saikrishna
ICFAI Foundation for Higher Education
(Deemed to be University)

CONTENTS

12.1   INTRODUCTION

12.1.1   PERMISSIONED BLOCKCHAIN

It is possible to architect, access, and manage a blockchain [1] in multiple ways. Some blockchains [1] need exceptional permissions for reading, transacting, writing, and accessing information about them. Such blockchains inherent configuration controls the interactions of the participant and defines their roles. Such blockchains are called permissioned blockchains.

A permissioned blockchain facilitates auxiliary security by providing an access control layer to the blockchain network, which permits only well-defined identities to transact with the distributed ledger in the network, without making data public. Permissioned blockchains permit different levels of permissions to be designated to its users; therefore, access can be restricted, sufficing confidentiality. A participant may join the network without any permission but may need authorization for transacting within the network. Hence such blockchains are favored by individuals who within the blockchain need to define security, identity, and role. Authorized blockchains are common among companies and businesses at the industry level, for which security, identity, and role management are relevant. A few advantages of a permissioned blockchain are listed below:

- **Closed Environment:** The enterprise controls the resources and access to the blockchain. Every transaction made within the network is isolated to itself. Hence, it ensures security and privacy up to an extent.
- **Transactions:** When nodes are distributed locally but also have much less nodes to participate in the network, the performance is faster ensuring that the transactions take less time to execute [2].
- **Scalability:** The ability to add nodes and services with ease can provide a great advantage to an enterprise.
- **Transparency and Accountability:** As all the participants in the network are well defined, every interaction made with the distributed ledger is traceable back to its provenance, and person who made a transaction is accountable for that particular transaction.
- As permissioned blockchains have their advantages, according to the industry aspect, it also has some disadvantages:
- **Conspiracy:** If in a permissioned blockchain, a group of members within the network collude to change the information in that blockchain, they will destroy the network's credibility.
- **Regulations:** As blockchains which are permissioned, operated by a group of enterprises and businesses, they must agree and stick to the regulations they implemented in the network. The threat of these regulations may not impact businesses that already comply with existing laws, but it may prevent new businesses from joining these permissioned blockchain networks.

12.2 PERMISSIONED BLOCKCHAIN FRAMEWORKS

As permissioned blockchains attracted many enterprises to explore the benefits of what it can offer, few frameworks gained popularity providing platforms. Some of them are Hyperledger Fabric [1], Quorum, Ethereum, and Corda [3,4].

TABLE 12.1
Differences between a Permissionless and Permissioned Blockchain

Category	Permissioned	Permission Less
Anonymity	Well-defined identities	Anonymous, sometimes partially anonymous
Privacy	Private membership	Open network
Ownership	Managed by a group of predefined identities	No one owns the network
Decentralization nature	Partially decentralized	Truly decentralized
Cost	Cost-effective	Not so cost-effective
Security (relating to anonymity)	Less secure	More secure

12.2.1 ETHEREUM

Ethereum [3,5], which is known as the most popular Blockchain platform after Bitcoin, is a decentralized platform that allows the creation of blockchain-based DApps (decentralized applications) using Ethereum smart contracts. Smart contracts [6] are self-executing digital contracts, which execute precisely as planned without any chance of lay-off, fraud, or intervention from third parties. Ethereum provides both public and permissioned configurations. Ethereum's main network [3,5] is an open blockchain network where anyone can enter the network and take part in the process of mining. As Ethereum also offers private blockchain configuration, it uses PoA (Proof of Authority) as a consensus mechanism. PoA doesn't need mining process, it requires only few authority nodes to achieve consensus. Hence, Ethereum does provide an option of private network but it doesn't offer a complete permissioned configuration required for an industry-level application.

12.2.2 QUORUM

Quorum [2] is an enterprise-focused offers decentralized platform that allows us to build permissioned blockchain-based DApps on top of it. Go Ethereum is a Golang-based implementation of the Ethereum protocol. Quorum, a hard fork of Go Ethereum is an enterprise-focused digital ledger technology, which offers a decentralized platform allowing us to build permissioned blockchain-based decentralized applications. Changes to the protocols and structures of a blockchain network are called forks [7]. There are two types of forks, namely soft fork [7] and hard fork [7]. Changes are backward compatible with nodes not patched for a soft fork [7]. Changes are not backward compatible for a hard fork [7] because the un-updated nodes would reject the blocks after the changes. Ethereum's large developer community and its privacy configurations drive a large number of enterprises to use Quorum for building blockchains. Microsoft Azure provides BaaS (Blockchain-as-Service) to easily build your own blockchain.

12.2.3 CORDA

Corda offers a platform to build your own permissioned distributed ledger technology-based applications. Corda is a company of R3, a firm that works to build Corda with several banks, financial institutions, regulators, several business associations, service-based businesses, and several technology firms. The idea of R3's Cordais to provide a distributed, trusted ledger for financial transactions. In Corda, smart contracts are known as CorDapps. They are not like smart contracts of other platforms; they don't have a state. Their purpose is to just validate if the outputs produced from the inputs are correct. Corda supports pluggable RDBMS (Relational database management system) to store smart contracts data. Multisignature support is additionally given by the framework that permits multiple nodes to sign a group action.

One of the major drawbacks of Corda is that as there is no global broadcasting, each peer (node) has to maintain its own backup and failover redundancy in a traditional way as there is no redundancy built into the network. A node will store transactions and retry sending the messages to the recipient until the recipient has successfully received it, as all transactions are not broadcasted to all parties in the network. Once the messages are received, the sender has no more responsibility.

12.2.4 HYPERLEDGER FABRIC

Hyperledger [1] is an open-source project of blockchains and related software released by the Linux Foundation in December 2015, which is backed by IBM. Fabric is the most popular project under Hyperledger. IBM is the top contributor to the project till date. Hyperledger Fabric [1] is a distributed ledger technology platform of enterprise grade level, which allows building permissioned blockchains. Where Hyperledger Fabric varies from other blockchain networks, it is personal as well as allowed. Instead of open permissionless system allowing unknown identities to participate in the network (requiring protocols such as "proof of work"(PoW) to verify transactions and protect the network), members of a Hyperledger Fabric network join through a trusted Membership Service Provider (MSP) [8].

Fabric also offers various pluggable configurations. Data on the ledger can be stored in multiple formats with pluggable consensus mechanisms and MSPs. Hyperledger Fabric introduces a concept called channels. Channels allow a group of identities from the same network to create a separate ledger for transactions, i.e., a separate subset blockchain of main blockchain. In this chapter, we are going to explore Hyperledger Fabric, its components and concepts, and a walkthrough for setting up a Fabric network. In the next section, we will be discussing about Fabric architecture.

12.3 FABRIC ARCHITECTURE

In this segment, we are going to discuss about Hyperledger Fabric's architecture its concepts and components which constitute a network.

12.3.1 FABRIC OVERVIEW

Hyperledger Fabric [1] is a modular and flexible open-source system designed to develop, deploy, and manage authorized (permissioned) blockchains. Traditional blockchains use active replication, i.e., a consensus process that initially orders the transactions and transmits them to all peers. Second, the transactions are executed sequentially by each peer. This needs all peers to be deterministic in conducting each transaction and all transactions, resulting in a lack of overall system performance. This is known as the architecture of order-execution. This architecture can be found in virtually all existing blockchain systems such as Ethereum (with PoW-based consensus) that are open to permitted platforms. Permissioned blockchains did suffer from some limitations with the traditional architecture. Consensus mechanism washard-coded, transactions need to be deterministic, which can be programmatically difficult to guarantee.

To overcome these drawbacks of existing permissioned blockchains, Hyperledger Fabric introduces a new architecture which they called execute-order-validate architecture [8]. In this architecture, transaction flow starts with the execution of the transactions, which are executed by a subset of peers (endorsing peers) in the network for checking its correctness, which can be considered as a transaction validation in other blockchains. Later, these transactions are ordered into a block using a consensus mechanism. Finally, transaction validation is done using an application-specific endorsement policy by all the peers before committing the transaction to the ledger.

With its new approach of execute-order-validate architecture, Hyperledger Fabric is much resilient, flexible, and scalable with enhanced performance and confidentiality over other existing blockchains with order-execute architecture. Fabric offers a highly configurable architecture. One of the configurational options is pluggable consensus. It can be considered as one of the major differentiating features of Hyperledger Fabric from other platforms. It drives the platform to be much effective and customized to fit diverse enterprise use cases When an enterprise with a trusted authority is considered, a simple crash fault tolerant (CFT) [4] consensus is much efficient over Byzantine Fault Tolerant (BFT) to avoid unnecessary drag on performance and throughput. Fabric doesn't need a native cryptocurrency for running consensus mechanisms like mining and to execute smart contract. The unneeded of cryptocurrency and costly cryptographic mining method will reduce the implementation and maintenance cost

Some other modular components of Fabric are pluggable ordering service. Ordering service is responsible for ordering the transactions that occurred in a given time interval. Pluggable membership service provider is responsible for identification and verification of the members transacting in the network with their cryptographic identities. Endorsement and validation policies are also pluggable.

This kind of modular design and features grabs the attention of enterprises convincing them to try and build appropriate, efficient permissioned blockchain systems to a wide range of industry use cases which include finance, banking, supply chain, and many more. The application level flow will be explained in the upcoming sections, and in the below section, we will discuss some major components and concepts of Hyperledger Fabric.

12.3.2 HYPERLEDGER FABRIC NETWORK CONCEPTS AND COMPONENTS

Hyperledger Fabric [9] network consists of several architectural components and concepts associated with them. A network consists of several organizations known as members. Through connecting the MSP to the network, an individual enters a network. Each MSP has a Certificate Authority (CA) that provides their members with cryptographic certificates. Each member may have more than one peer participating in the blockchain network representing its organization. These peers in the network of different organizations or the same organization form a channel to share a ledger. Before diving into how things work on the network, we need to get familiar with some keywords, concepts, and components described below:

12.3.2.1 Organization

Organizations are also known as members. The provenance for a network begins by joining organizations to the network. A blockchain network administrator who creates a network can invite members into the network. Initially, an organization joins the network when its MSP is added to the network. These members participate in the network representing their organizations. Organization manages its members under a single MSP.

12.3.2.2   Consortium

A consortium is a group on the blockchain network of non-ordering organizations. These are the channel-forming and entering organizations and their own peers. While there may be several consortia for a blockchain network, most blockchain networks have one consortium. All entities introduced to the network must be part of a partnership at the time of the creation of the channel. Nevertheless, an entity may be linked to an existing network that is not specified in a consortium.

12.3.2.3   Membership Service Provider (MSP)

MSP is like an abstraction layer to all cryptographic mechanisms and procedures behind issuing, validating, and authenticating certificates, which are issued by the Certification Authority to access a network. MSP can use one or more Certificate *Authorities*; these interfaces are configurable. The Fabric-CA API (application programming interface) is the default interface used for the MSP. This is another modular feature of Hyperledger Fabric. An MSP can define its own regulations attributing their identities.

MSP's responsibility is to validate and authenticate an identity, i.e., the generation and verification of transaction signature. More than one MSP can govern a Fabric network. It ensures modularity and interoperability of membership processes across various membership standards and architectures. Fabric supports various certificate architectures that facilitate the use of interfaces from the External CA that allow organizations to bring in their own trusted authorities.

12.3.2.4   Certificate Authority (CA's)

An individual or node may participate in the network through a digital identity provided for it by a system trusted authority. Digital identities or simply identities have, in the most common case, the form of cryptographically authenticated digital certificates that follow the X.509 standard and are provided by a CA.

Certificates are issued by a Certification Authority to various actors. The CA signs these certificates electronically and links the actor to the actor's public key (and optionally to a specific property list). Therefore, if you trust the CA (and know its public key), you can assume that a particular actor is bound to the public key included in the certificate and retains the attributes included by validating the CA signature of the actor.

12.3.2.5   Channels

Channel is one of the most important concepts of Fabric network. Channels allow the same network to be used by organizations while maintaining complete isolation among multiple blockchains. It's like a big blockchain branch. Only the channel participants on whom the transaction was performed can see the transaction's details.

Figure 12.1 depicts how Fabric supports multiple channels in a single main network. Each channel will have its own ledger completely isolated from one another, including the main network. Transactions within a channel will not affect transactions in another channel and same goes to ledgers

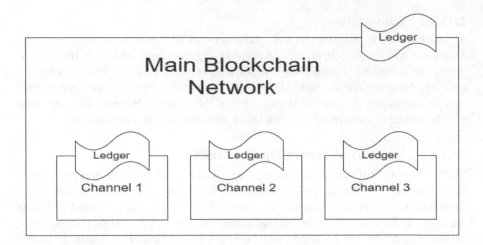

FIGURE 12.1 Main blockchain network.

12.3.2.6 Peers

Peers are a network body that maintain a ledger as shown in Figure 12.2 and executes chain code to perform read/write operations in the network. Members own and retain peers. A peer can host more than one ledger and chaincode to help create more robust layout of the process. While all peers are the same, depending on how the network is designed, they may play multiple roles.

Committing peer: Every peer node in a channel is a peer to commit. It receives blocks of transactions created, which are subsequently validated before they are committed and attached to the ledger's peer local copy.

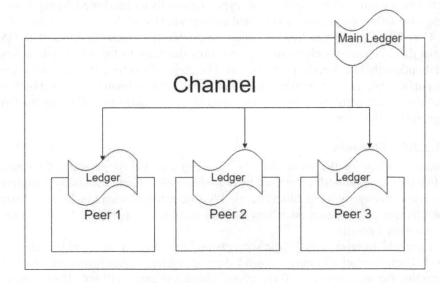

FIGURE 12.2 Channel–peer architecture.

Endorsing Peers: Endorsing peers are those peers who simulate transactions in isolated chaincode containers and prepare a transaction proposal based on smart contract results. However, in order to be an endorsing peer, a client application must use the smart contract on the peer to generate a digitally signed response to the transaction.

Anchor Peer: If a peer in another company wants to connect with a peer, it can use one of the channel setup anchor peers. An organization will have zero or additional peers outlined for it which act like anchors, and many different cross-organization interactions can be supported by these peers. They use gossip protocols to communicate within the network.

12.3.2.7   Peer Gossip

Peers use gossip to transmit ledger and channel information. Gossip communication is continuous and every peer on a channel receives from multiple peers constantly current and reliable ledger information. A message or transaction, which is being communicated in the network, is signed to prevent unwanted targets from transmitting the messages. Peers affected by delays, network partitions, or other triggers that lead to missed blocks must finally be synchronized to the present ledger state by communicating with its neighboring peers.

The three functions performed during this communication process are:

Step 1: Managing peer discovery and channel membership by continuous recognition of active member peers and gradually detection of off-line peers.

Step 2: Disseminates leader information on a network across all peers. With the rest of the channel, each peer with data out of sync detects and synchronizes the missing blocks by copying the correct data.

Step 3: By allowing peer-to-peer transmission of ledger information to be modified, bring newly linked peers to speed.

Peers receive messages from other peers on the channel, and these messages are forwarded to some random peers. One peer act as leader on the network, takes the information from the ordering system for the distribution of new blocks, and initiates gossip dissemination in his own company to peers.

Leader Peer: When an organization has multiple peers in a channel, a leader peer is a node that assumes responsibility for distributing transactions from the orderer to the organization's other committing peers. A Leader peer may opt to participate either in static, where peer is manually configured or dynamic, leadership choice where peers mutually execute a procedure to elect one.

12.3.2.8   World State and Databases

Apart from a peer maintain its own ledger, Fabric introduces a concept of world state. A world state maintains the latest set of values, i.e., current state, while a ledger maintains transaction log which includes the history of all transactions and data alterations.

The current state data represents the new values for all assets in the ledger. Because all transactions committed on the network are reflected in the current state, it is sometimes referred to as a world state. A set of key value pairs can be the world state. Chaincode invocations (smart contracts) are conducting transactions against current state information. To make these chaincode interactions extremely efficient, the latest key-value pairs for each asset are stored in a state database.

The state database is simply an indexed view of the transactions committed. Thus, it can be regenerated from the string at any time. The state database will automatically accept new transactions

a. LevelDB is Hyperledger Fabric's default key/value state repository it stores key-value pairs.
b. CouchDB is a LevelDB substitute. In comparison to LevelDB, CouchDB stores data in JavaScript Object Notation (JSON) format. CouchDB supports main, composite, key scope, and full data-rich queries.

12.3.2.9 Chaincode

The popularly used term, smart contract, is called chaincode in Hyperledger Fabric. Chaincode enables participants to transact and update ledger and the world state. Chaincode allows a properly approved user to download and install chaincode on the peers of a network. Chaincode only rests within the peer. Using client-side software that interacts with a peer in the network, end users can invoke chaincode. You can install same chaincode on multiple channels. A peer can hold multiple chaincodes.

12.3.2.10 Ordering Service

Most of the distributed blockchains, which are permissionless, permit anyone to participate in the network consensus, where the transactions are bundled and ordered into block. Hyperledger Fabric works differently as explained in the previous sections, that it introduced a new architecture called order-execute-validate. It contains a module known as the ordering node also known as orderer, *which performs ordering of transactions*. A Fabric ordering service accepts transactions that have been endorsed, orders them into a block, and delivers the blocks to committing peers. It does not maintain a ledger or chaincode with the sole purpose of ordering transactions into a block and distributing the block to all peers in the network. This is considered an ordering service. Because Fabric relies on deterministic algorithms of consensus, it is difficult to fork the ledger. Hyperledger Fabric offers three ordering mechanisms:

a. SOLO: It is most widely used by programmers working with Hyperledger Fabric networks that are the Hyperledger Fabric ordering process. SOLO requires a single command node.
b. Kafka: Recommended for production use is the Hyperledger Fabric ordering system. This mechanism uses Apache Kafka, an open-source stream processing platform that provides a unified low-latency, high-throughput platform for handling data feeds in real time. The data consists of approved transactions and RW sets in this case. A crash-tolerant solution for ordering is given by the Kafka method.
c. SBFT: It stands for Simplified Byzantine Fault Tolerance. The ordering system is both crash-tolerant and BFT, which ensures that even in the presence of malicious or faulty nodes it can reach agreement. This method has not yet been adopted by the Hyperledger Fabric team, as it is under development.

Orderers also maintain a list of channel-building organizations also called as a consortium. This list is kept in orderer system channel. By definition, only the orderer administrator can only change this list and the channel on which it resides.

12.3.2.11 Endorsing Policy

Policy on endorsement determines the smallest collection of organizations needed to support a transaction to be legitimate. To endorse, the endorsing peer of an entity needs to run the transaction-related smart contract and sign its result.

When transactions are ordered and broadcasted to the committing peers, they must ensure that each endorsed transaction in the block complies with the endorsing policy defined in the network. If a transaction doesn't satisfy the policy when it is marked as invalid and world state remains unchanged with its values.

12.3.2.12 Private Data Collections

In Hyperledger Fabric [1], data can be privatized, i.e., access to particular data can be restricted within the same channel. Fabric calls this as private data collection. Unauthorized entities, as confirmation of the transaction data, will have a hash of private data on the network ledger. It will store this private in a different database called as sideDB [8] as shown in Figure 12.3 which exists in the same peer, if configured.

If a peer is authorized to have a private data collection the private state exists within the peer. This collection is isolated from neighboring peers and only the peer itself has access to it.

12.3.3 Transaction Flow

In this segment, we are going to depict how a transaction is initiated, endorsed, ordered, and committed to the ledger in Hyperledger Fabric network [1].

The transaction flows into five steps.

Step 1: Transaction Proposal

Transaction life cycle in the Hyperledger Fabric network starts with client applications submitting transaction proposals or, in other words, providing a transaction to endorse peers. A transaction is first invoked by the client application.

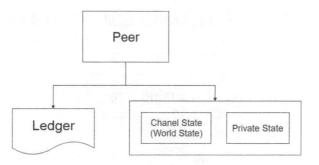

FIGURE 12.3 Private data architecture.

Figure 12.4 depicts how an end-user, i.e., customer uses software applications to connect with the blockchain network leveraging Fabric Software Development Kit (SDK) and sends its transaction proposal to the required endorsing peers in the network.

Step 2: Endorsing and Generating RW Sets

Endorsing peers by invoking the respective chaincode at the request of the client user, executes the proposed transaction and produces the output. Endorsing peer then gathers RW sets (read set and write set) information. Read set is the values of the current state of the world without changing it with the transaction proposed. Write set is the current state modified values prior to the execution of the proposed transaction. Such improvements, however, are not reflected at this point in the ledger.

Figure 12.5 depicts how all endorsing peers sign the created RW sets cryptographically and return them to the client request. Such RW sets are also used to complete the transaction by the client request. Depending on the endorsement policy,

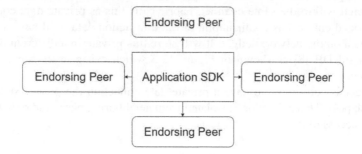

FIGURE 12.4   Transaction proposal.

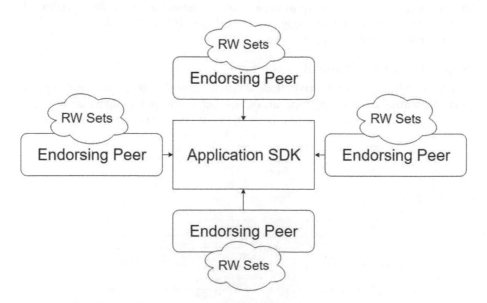

FIGURE 12.5   Endorsing and generating RW sets.

the number of peers required to sign a transaction. When deploying the chaincode, this endorsement policy is defined. Therefore, different channels can have different endorsement policies.

Transaction Flow (Step 3): Ordering

After receiving endorsed transactions and signed RW sets, the client application submits the endorsed transaction and RW sets to the ordering service. Ordering service collects all the transactions made in a particular time interval, orders all the transactions of that particular channel, and groups them all into a block. The ordering service can be a group of orderers, does not process transactions or maintains a shared ledger. The ordering service only specifies the order in which the transactions will be committed to the ledger.

Figure 12.6 depicts how a user/client sends endorsed transactions received from endorsing peers and sends them to ordering service.

Transaction Flow (Step 4): Broadcasting

The ordering system takes the accepted transactions and RW sets, ordering this information into a block, and distributing the block to all peers who commit.

The ordering system, which consists of a group of orderers, does not process transactions, smart contracts, or the distributed ledger is maintained. The ordering system recognizes the transactions that have been approved and determines the order in which those transactions must be appended ledger and world state in upcoming steps.

Figure 12.7 depicts how an ordering service orders transactions into a block and broadcasts the block to the other peers in the network.

Step 5: Committing and Ledger Updating

Committing peers validate every transaction in the block separately. Considering the Read Write sets, the peer's checks if the current world state and the read set are same as shown in Figure 12.8. Then validates the transaction by verifying all the endorsements required to satisfy the endorsement policy.

After the validation process, shown in Figure 12.8, the ledger is updated. Ledger updating by appending the block is done synchronously by all the peers. Invalid transactions detected are marked as invalid, still present in the block but do not affect the world state. Committing peers are the one updating and maintaining the shared ledger. They also notify the client application about the success or failure of

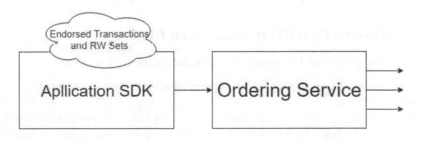

FIGURE 12.6 Application SDK to ordering service.

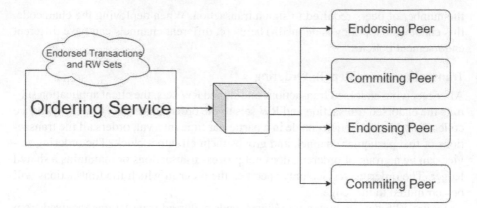

FIGURE 12.7 Ordering service to peers in network.

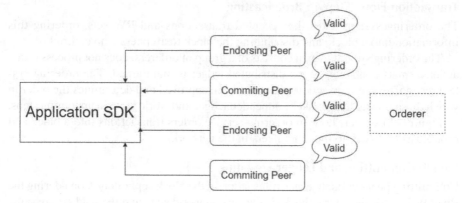

FIGURE 12.8 Committing.

the transaction updating to the ledger. Lastly, the committing peers asynchronously notify the client application of the success or failure of the transaction. Applications will be notified by each committing peer.

Apart from endorsement of transactions, validity, and versioning the system, identity verifications run in the background during each step of the transaction flow. Transaction inputs and outputs are repeatedly authenticated, i.e., signing and verifying as a transaction proposal passes through the different components in the network.

12.4 HANDS-ON WITH HYPERLEDGER FABRIC

12.4.1 Demo of the Hyperledger Fabric Network

In this segment, we will address the network that will be built in hands-on with Hyperledger Fabric.

This section will give you the template that will help you build a Hyperledger Fabric with three organizations with two channels and one orderer, every organization will have one peer and the peer will join the channels, the peer will also have a

CouchDB container to save the world state of the ledgers. The Fabric network will have two channels. Peers of Org-3 and Org-2 can join channel channel1 and peers of Org-1 and Org-3 can join channel channel2. We will also build the chaincode (smart contract) for the channel1 and channel2, and build REST API using Fabric SDK to interact with the Fabric network

Figure 12.9 depicts the architecture of a Fabric network. We will first go through the prerequisites and next see the configurations required for the organization-specific container definitions for docker-compose file, we will look into the orderer configuration for the docker-compose file. Then we will look into configtx.yaml configuration file, crypto-config.yaml configuration file. After finishing with the configuration files, we will develop a basic chaincode for both the channels. Next, we will see the various commands used to create the certificates using the crypto-config configuration file, generate the genesis block, generating channel transaction, generating anchor peer channel transaction file, channel creation, channel join, anchor peer update, chaincode installation, chaincode instantiation. And finally, we will write the REST API using Fabric Node SDK to interact with the Hyperledger Fabric network.

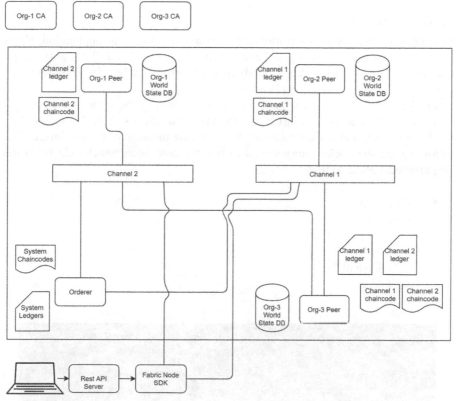

FIGURE 12.9 Network architecture.

12.4.2 Prerequisites

Before we begin developing the Hyperledger Fabric network, consider the following software that should be running in your machine to build the network successfully.

- Docker
- Docker-compose
- Node.js

Figure 12.10 displays command to download and execute bash script to download and extract the platform-specific binaries.

Determine the location where you want to save the Fabric-samples repository, cloned when executing the above command. The command will perform the following:

- Clones the Fabric-samples repository
- Download the platform specific binaries and config files into the bin and config directories of the Fabric-samples.
- Download the Fabric docker images. You can check the images downloaded by using the command *"docker images Hyperledger/fabric*"*.

12.4.2.1 Docker

Docker is a containerization platform that wraps up all of your application, all its necessary dependencies into a package called container. This will not only ensure that the application will work seamlessly in any environment but also provides better delivery of production ready application.

Docker containers run on the same machine sharing the same operating system kernel, this makes them faster than the virtual machines. A docker image does not have a state, and docker container is a runtime instance of docker image. The following are the docker images used to run the containers which help build the Hyperledger Fabric network.

- Fabric-peer
- Fabric-kafka
- Fabric-zookeeper
- Fabric-orderer
- Fabric-CouchDB
- Fabric-CA

```
1 curl -sSL http://bit.ly/2ysbOFE | bash -s
```

FIGURE 12.10   Command to download binaries.

The above-mentioned docker images are pulled from the hub.docker.com, a service for finding and sharing docker images. The docker images are customized for each organization by passing the environmental variables when starting the containers.

Docker-compose is a tool used for running multiple containers as a single service. We use YAML configuration file to run our application services. To build our HyperledgerFabric network, we will run multiple Fabric-peer, Fabric-kafka, Fabric-zookeeper, Fabric-orderer, Fabric-CouchDB, Fabric-CA. And to run the network as a whole, we will use docker-compose.

As mentioned before, in this hand-on section we are going build a Hyperledger Fabric network with three organizations. An organization needs a CA, at least one peer, and one CouchDB for each peer. The CA is used for managing user certifications like user certification, user enrolment, and user revocation. The peer is a node that joins the channels and stores the transactions in the ledger. A peer can join multiple channels but each channel will have different storage, so an organization doesn't need to worry about the confidential information shared with specific peers with channels. A CouchDB is used to store all the ledgers world state that the peer holds.

There are three configuration files where we will be adding the organization details.

- docker-compose.yaml
- Crypto-config.yaml
- configtx.yaml

In the docker-compose file, we define the containers required for the organization to run in the Hyperledger Fabric network. The organization details in the crypto-config file are used to generate the certificate materials for the organization. To generate the channel-artifacts, configtx.yaml file is used.

The crypto-config.yaml and configtx.yaml configuration files are explained later in the section. Here, we will see how to setup the docker-compose.yaml file for an organization. The below configurations are template for running the containers for organization.

12.4.3 Docker-Compose Services Configuration

The first line ca.org1.hnetwork.com in Figure 12.11 is the service name in the docker-compose file, the second line gives the container name when the compose is started, and the third line is the image used to run the container.

The fifth and sixth lines are the environmental variables passed to the container. The eighth line is about the ports; the left side of the colon is the port exposed to the system, and the right is the port from the container, when you create more than one organization which needs more than one CA to run the network, we need to change the value of the left side to the colon to some other port. Because the users cannot use the same port to expose to the system by two containers.

The ninth line, the command to run the container is defined; the next line consists of the volume mounting. To save the data or read the data from the system to the container, the volumes are used. For the CA container, we pass the organization CA certificates.

```
1 ca.org1.hnetwork.com:
2     container_name: ca.org1.hnetwork.com
3     image: Hyperledger/fabric-ca
4     environment:
5       - FABRIC_CA_HOME=/etc/Hyperledger/fabric-ca-server
6       - FABRIC_CA_SERVER_CA_NAME=ca.org1.hnetwork.com
7     ports:
8       - "7054:7054"
9     command: sh -c 'fabric-ca-server start --ca.certfile /etc/Hyperledger/fabric-ca-
  server-config/ca.org1.hnetwork.com-cert.pem --ca.keyfile /etc/Hyperledger/fabric-ca-
  server-config/*_sk -b admin:adminpw -d'
10    volumes:
11      - ./crypto-
  config/peerOrganizations/org1.hnetwork.com/ca/:/etc/Hyperledger/fabric-ca-server-config

12    networks:
13      - basic
14
```

FIGURE 12.11   CA docker service definition in the docker-compose file.

And the last line defines the network that the service is used to run on.

The first line, peer0.org1.hnetwork.com as shown in Figure 12.12, is the service name in the docker-compose file, the second line gives the container name when the compose is started, and the third line is the image used to run the container.

```
1 peer0.org1.hnetwork.com:
2     container_name: peer0.org1.hnetwork.com
3     image: Hyperledger/fabric-peer
4     environment:
5       - CORE_VM_ENDPOINT=unix:///host/var/run/docker.sock
6       - CORE_PEER_ID=peer0.org1.hnetwork.com
7       - FABRIC_LOGGING_SPEC=debug
8       - CORE_CHAINCODE_LOGGING_LEVEL=DEBUG
9       - CORE_PEER_LOCALMSPID=Org1MSP
10      - CORE_PEER_MSPCONFIGPATH=/etc/Hyperledger/msp/peer/
11      - CORE_PEER_ADDRESS=peer0.org1.hnetwork.com:7051
12      - CORE_VM_DOCKER_HOSTCONFIG_NETWORKMODE=${COMPOSE_PROJECT_NAME}_basic
13      - CORE_LEDGER_STATE_STATEDATABASE=CouchDB
14      - CORE_LEDGER_STATE_COUCHDBCONFIG_COUCHDBADDRESS=couchdb:5984
15      - CORE_LEDGER_STATE_COUCHDBCONFIG_USERNAME=
16      - CORE_LEDGER_STATE_COUCHDBCONFIG_PASSWORD=
17      - CORE_PEER_GOSSIP_BOOTSTRAP=peer0.org1.hnetwork.com:7051
18      - CORE_PEER_GOSSIP_EXTERNALENDPOINT=peer0.org1.hnetwork.com:7051
19    working_dir: /opt/gopath/src/github.com/Hyperledger/fabric
20    command: peer node start
21    ports:
22      - 7051:7051
23      - 7053:7053
24    volumes:
25      - /var/run/:/host/var/run/-
26      - ../chaincode:/etc/Hyperledger/chaincode
27      - ./crypto-
  config/peerOrganizations/org1.hnetwork.com/peers/peer0.org1.hnetwork.com/msp:/etc/Hyper
  ledger/msp/peer
28      - ./crypto-
  config/peerOrganizations/org1.hnetwork.com/users:/etc/Hyperledger/msp/users
29      - ./config:/etc/Hyperledger/configtx
30    depends_on:
31      - orderer.hnetwork.com
32      - couchdb
33    networks:
34      - basic
35
```

FIGURE 12.12   Peer docker service definition in a docker-compose file.

The lines from fifth to sixteenth are the environmental variables passed to the peer container. The environment variable value of the seventh line "FABRIC_ LOGGINH_SPEC=debug" can be any of FATAL|PANIC|ERROR|WARNING|INF O|DEBUG. The fifteenth and sixteenth lines are the credentials required for the peer to connect to the CouchDB instance.

The working directory sets the working directory path for the peer container. The next line, the command to run the container is defined, and the exposing ports are defined in the next line.

The depends_on array in the 30th line is used to define what containers should be up and running before this starts. And the last line defines the bridge-network that the service is using to run on.

CouchDB Docker service definition for an organization. The first line is the service name, second line as shown in Figure 12.13 tells about the container name, and next line about the image used for the service. The environmental variables passed are the username and password for the CouchDB instance. The ports will be exposed to the host and the final line defines the bridge-network that the service is using to run.

12.4.3.1   Orderer

Hyperledger Fabric provides few ordering mechanisms; some of them are SOLO, Kafka, RAFT.

Developers playing with the Fabric use the SOLO ordering method. In the production stage, Kafka is the ordering system that could be used, it used the Apache Kafka. This system provides an ordering solution resistant of crash failure. RAFT ordering system offers both crash tolerance and Byzantine Fault Tolerance, which ensures that even if there are unstable or malicious nodes, agreement can be reached.

We will be using the Kafka ordering system in the Hyperledger Fabric network.

```
 1 couchdb0:
 2     container_name: couchdb0
 3     image: Hyperledger/fabric-couchdb
 4     environment:
 5         - COUCHDB_USER=
 6         - COUCHDB_PASSWORD=
 7     ports:
 8         - 5984:5984
 9     networks:
10         - basic
11
```

FIGURE 12.13   CouchDB docker service definition in a docker-compose file.

The crypto-config.yaml and configtx.yaml configuration files are explained later in this section. Here, we will see how to setup the docker-compose.yaml file for an organization. The below configurations are template for running the containers for organization.

Figure 12.14 displays Orderer Docker service definition in a docker-compose file.

The first line as shown in Figure 12.14 is the service name, and the next two lines are the container name and the image used by the container. The fourth line sets the working directory for the container and the next line is the command used to start the container. The environment variables define what type of ordering service the orderer must use and the location of the genesis block and some other variables.

The port mapping is 7050:7050, the left value to colon is the exposed port to the host, and the right one is the port that the container maps to. The persistent volumes mapped are the channel artifacts folder and the MSP certificates of the orderer. The orderer depends on the kafka brokers. The last line defines the bridge-network that the service is used to run on.

In Figure 12.15, the first line is the service name, and then follows the container name and the image that is used to run the container. The environment variables defined include KAFKA_DEFAULT_REPLICATION_FACTOR which defines the default replication factor and KAFKA_MIN_INSYNC_REPLICAS and how many minimum Kafka must be in sync (Figure 12.16).

```
1 orderer.fnetwork.com:
2     container_name: orderer.fnetwork.com
3     image: Hyperledger/fabric-orderer:latest
4     working_dir: /opt/gopath/src/github.com/Hyperledger/fabric
5     command: orderer
6     environment:
7         - CONFIGTX_ORDERER_ORDERERTYPE=kafka
8         - CONFIGTX_ORDERER_KAFKA_BROKERS=
  [kafka0:9092,kafka1:9092,kafka2:9092,kafka3:9092]
9         - ORDERER_KAFKA_RETRY_SHORTINTERVAL=1s
10        - ORDERER_KAFKA_RETRY_SHORTTOTAL=30s
11        - ORDERER_KAFKA_VERBOSE=true
12        - ORDERER_GENERAL_LOGLEVEL=debug
13        - ORDERER_GENERAL_LISTENADDRESS=0.0.0.0
14        - ORDERER_GENERAL_GENESISMETHOD=file
15        - ORDERER_GENERAL_GENESISFILE=/etc/Hyperledger/configtx/genesis.block
16        - ORDERER_GENERAL_LOCALMSPID=OrdererMSP
17        - ORDERER_GENERAL_LOCALMSPDIR=/etc/Hyperledger/msp/orderer/msp
18    ports:
19        - 7050:7050
20    volumes:
21        - ./channel-artifacts:/etc/Hyperledger/configtx
22        - ./crypto-
  config/ordererOrganizations/fnetwork.com/orderers/orderer.fnetwork.com/msp:/etc/Hyperledger/msp/orderer/msp
23    depends_on:
24        - kafka0
25        - kafka1
26        - kafka2
27        - kafka3
28    networks:
29        - basic
```

FIGURE 12.14   Orderer docker service definition in a docker-compose file.

```
 1 kafka0:
 2     container_name: kafka0
 3     image: Hyperledger/fabric-kafka
 4     ports:
 5        - 2181
 6     environment:
 7        - KAFKA_LOG_RETENTION_MS=-1
 8        - KAFKA_UNCLEAN_LEADER_ELECTION_ENABLE=false
 9        - KAFKA_DEFAULT_REPLICATION_FACTOR=3
10        - KAFKA_MIN_INSYNC_REPLICAS=2
11        - KAFKA_BROKER_ID=0
12        - KAFKA_ZOOKEEPER_CONNECT=zookeeper0:2181,zookeeper1:2181,zookeeper2:2181
13        - KAFKA_MESSAGE_MAX_BYTES=103809024
14        - KAFKA_REPLICA_FETCH_MAX_BYTES=103809024
15        - KAFKA_REPLICA_FETCH_RESPONSE_MAX_BYTES=103809024
16     depends_on:
17        - zookeeper0
18        - zookeeper1
19        - zookeeper2
20     networks:
21        - basic
```

FIGURE 12.15   Kafka docker service definition in docker-compose file.

```
 1 zookeeper0:
 2     container_name: zookeeper0
 3     image: Hyperledger/fabric-zookeeper
 4     ports:
 5        - 2181
 6        - 2888
 7        - 3888
 8     environment:
 9        - ZOO_MY_ID=1
10        - ZOO_SERVERS=server.1=zookeeper0:2888:3888 server.2=zookeeper1:2888:3888
   server.3=zookeeper2:2888:3888
11     networks:
12        - basic
```

FIGURE 12.16   Zookeeper docker service definition in docker-compose file.

12.4.4   CRYPTO-CONFIG.YAML CONFIGURATION

The crypto config configuration file is used to generate the certification files for the Hyperledger Fabric network with the help of cryptogen tool.

There are two objects, OrdererOrgs and PeerOrgs as shown in Figure 12.17, the ordererOrgs consists of the orderer details like the name of the orderer, Domain, and hostname. The PeerOrgs consists of the name of the organization, the number of peers, and the number users in addition to admin certificates.

Certification materials generation command using the cryptogen tool

Figure 12.18 displays the command of a cryptogen tool

Figure 12.19 displays the file structure of the crypto-config folder, which is generated after the crypto-config.

```
○ ○ ○

1 OrdererOrgs:
2   - Name: Orderer
3     Domain: fnetwork.com
4     Specs:
5       - Hostname: orderer
6 PeerOrgs:
7   - Name: org1
8     Domain: org1.fnetwork.com
9     Template:
10      Count: 1
11    Users:
12      Count: 3
13  - Name: org2
14    Domain: org2.fnetwork.com
15    Template:
16      Count: 1
17    Users:
18      Count: 3
19  - Name: org3
20    Domain: org3.fnetwork.com
21    Template:
22      Count: 1
23    Users:
24      Count: 3
```

FIGURE 12.17   Crypto config configuration file.

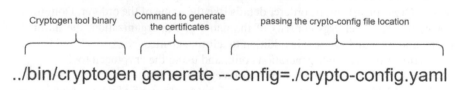

FIGURE 12.18   Cypto files generation command.

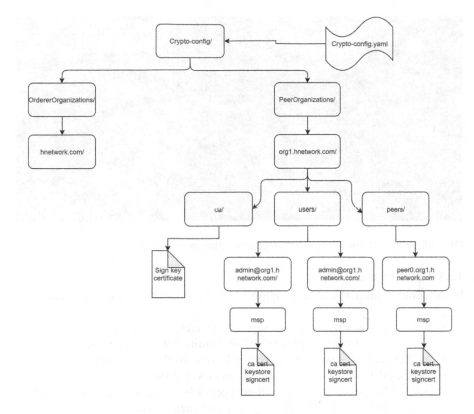

FIGURE 12.19   Crypto-config folder structure.

12.4.4.1   Configtx.yaml Configuration

Configtx.yaml configuration is used to generate the channel transactions, which are the metadata about the channel rather than the ledgers state data. Both the channel transactions and the normal transactions are stored on the same blockchain. The difference is that each channel transaction is stored in a special block that contains not only the incremental configuration update but the entire configuration as well. When a peer or client needs to know the channel's configuration, it only needs to get the latest channel transaction block.

Configtx.yaml defines which elements to use with your network and is passed to configtxgen in order to create genesis block and channel artifacts.

We can split the configtx.yaml file into five different blocks:

- Organizations block.
- Capabilities block.
- Applications block.
- Ordererblock.
- Profiles block.

Figure 12.20 depicts the defining organization configuration in configtx.yaml

```
1 - &org1
2     Name: org1
3     ID: org1MSP
4     MSPDir: crypto-config/peerOrganizations/org1.fnetwork.com/msp
5     Policies:
6       Readers:
7         Type: Signature
8         Rule: "OR('org1MSP.admin', 'org1MSP.peer', 'org1MSP.client', 'org1MSP.member')"
9       Writers:
10        Type: Signature
11        Rule: "OR('org1MSP.admin', 'org1MSP.client', 'org1MSP.member')"
12      Admins:
13        Type: Signature
14        Rule: "OR('org1MSP.admin', 'org1MSP.member')"
15    AdminPrincipal: Role.MEMBER
16    AnchorPeers:
17      - Host: peer0.org1.fnetwork.com
18        Port: 7051
```

FIGURE 12.20   Defining organization configuration in configtx.yaml.

12.4.4.2 Organizations

In the configtx.yaml as shown in Figure 12.20 the organization data represents

- Name is the key by which this org will be referenced in channel.
- ID is the key by which this org's MSP definition will be referenced.
- MSPDir contains about the path which contains the MSP configuration.
- Policies defines the set of policies at this level of the config tree.
- Anchor Peersis used to define all the peers a host location and port number, which can be used for cross-org gossip communication.

12.4.4.3 Capabilities

Since Hyperledger Fabric is a distributed system that will involve multiple organizations, there is a chance that different version of Fabric may exist in different nodes within the Fabric network and on channels in the Fabric network as shown in Figure 12.21.

- Channel: These capabilities apply to orderers and peers and must be supported by both.
- Orderer: These capabilities apply to the orderers and may be safely used with prior release peers.
- Application: These capabilities apply to the peer network and may be safely used with prior release orderers.

12.4.4.4 Channel Defaults

Channel Defaults defines the values that are to be encoded into a config transaction or genesis block for channel-related parameters as shown in Figure 12.22
Policies:

- *Readers*: Who may invoke the "Deliver: API.
- *Writers*: Who may invoke the "Broadcast" API.
- *Admins*: By default, who may modify elements at this config level.

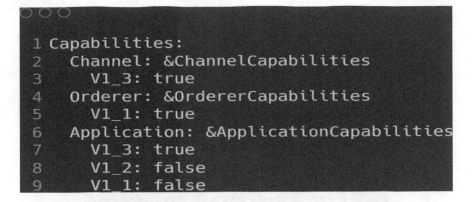

```
1 Capabilities:
2    Channel: &ChannelCapabilities
3       V1_3: true
4    Orderer: &OrdererCapabilities
5       V1_1: true
6    Application: &ApplicationCapabilities
7       V1_3: true
8       V1_2: false
9       V1_1: false
```

FIGURE 12.21   Capabilities configuration in configtx.yaml.

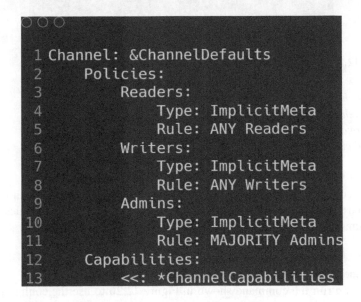

```
1 Channel: &ChannelDefaults
2        Policies:
3            Readers:
4                Type: ImplicitMeta
5                Rule: ANY Readers
6            Writers:
7                Type: ImplicitMeta
8                Rule: ANY Writers
9            Admins:
10               Type: ImplicitMeta
11               Rule: MAJORITY Admins
12        Capabilities:
13               <<: *ChannelCapabilities
```

FIGURE 12.22   Channel defaults in Configtx.yaml.

12.4.4.5   Application Defaults
Application Defaults defines the values that are to be encoded into a config transaction orgenesis block for application related parameters as shown in Figure 12.23.

12.4.4.6   Orderer Defaults
Orderer defaults defines the values that are to be encoded into a config transaction orgenesis block for orderer-related parameters as shown in Figure 12.24.

- *Orderer Type*: The type of ordering service used.
- *Addresses*: Used to be the list of orderer addresses that clients and peers could connect to.

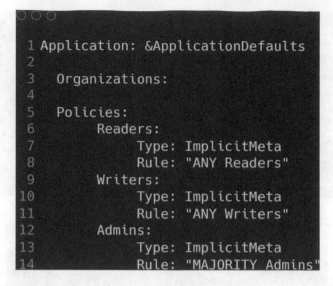

```
 1 Application: &ApplicationDefaults
 2
 3   Organizations:
 4
 5   Policies:
 6       Readers:
 7             Type: ImplicitMeta
 8             Rule: "ANY Readers"
 9       Writers:
10             Type: ImplicitMeta
11             Rule: "ANY Writers"
12       Admins:
13             Type: ImplicitMeta
14             Rule: "MAJORITY Admins"
```

FIGURE 12.23   Application defaults in configtx.yaml.

- *Batch Timeout*: The time to wait before a batch is produced.
- *Batch Size*: Controls the number of batched messages in a block.
- *Brokers*: A list of Kafka brokers linked to by the orderer.
- *Policies*: Defines the set of policies at this level of the config tree.

12.4.4.7   Profiles

Different configuration profiles can be encoded here for the configtxgen tool to be specified as parameters. To generate the orderer genesis block, the profiles defining consortiums are to be used (Figure 12.25).

12.4.4.8   Channel Profile

The channel profile describes the consortium for the channel.

Before we run the command shown in Figure 12.26 to use the configtxgen tool, we must create a channel-artifacts folder in the current working directory.

12.4.5   Commands

```
$ mkdir channel-artifacts
$ export CHANNEL_NAME=channel1
$ export CHANNEL_PROFILE=Channel1
```

Command to generate the orderer genesis block.

Figure 12.27 shows different command where the outputs are generates the block and saves it in the path specified.

After creating the genesis.block using profile OrdererGenesis, we will generate the channel transactions. Genesis block contains consortium information; basically

```
1 Orderer: &OrdererDefaults
2    OrdererType: kafka
3    Addresses:
4       - orderer.fNetwork.com:7050
5    BatchTimeout: 2s
6    BatchSize:
7       MaxMessageCount: 10
8       AbsoluteMaxBytes: 98 MB
9       PreferredMaxBytes: 512 KB
10   Policies:
11         Readers:
12               Type: ImplicitMeta
13               Rule: "ANY Readers"
14         Writers:
15               Type: ImplicitMeta
16               Rule: "ANY Writers"
17         Admins:
18               Type: ImplicitMeta
19               Rule: "MAJORITY Admins"
20         BlockValidation:
21               Type: ImplicitMeta
22               Rule: "ANY Writers"
23      Kafka:
24        Brokers:
25        - kafka0:9092
26        - kafka1:9092
27        - kafka2:9092
28        - kafka3:9092
29      Organizations:
```

FIGURE 12.24 Orderer defaults configuration in Configtx.yaml.

it contains all certificates of organizations and CA's. Therefore, allowing to initialize channels MSPs and use those root CAs certificates to validate Access Control Lists (ACLs), endorsements, and clients signatures.

Figure 12.28 shows command which uses the channel profile from the configtx. yaml and creates the channel transaction file. This file is submitted to the system channel, and it includes configuration required for formation of the new channel. Such as channel policies and members of the channel consortium which by the way have to be a subset of the consortium defined within genesis block of the system channel. This channel transaction is required by any peer who wants to join the channel.

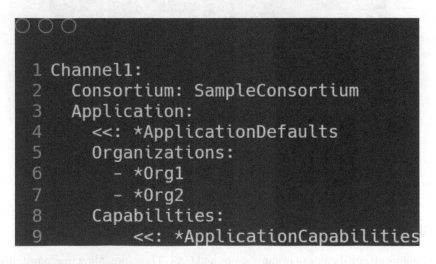

```
1  OrdererGenesis:
2    Capabilities:
3      <<: *ChannelCapabilities
4    Orderer:
5      <<: *OrdererDefaults
6      Organizations:
7        - *OrdererOrg
8      Capabilities:
9        <<: *OrdererCapabilities
10   Consortiums:
11     SampleConsortium:
12       Organizations:
13         - *Org1
14         - *Org2
15         - *Org3
```

FIGURE 12.25   Orderer genesis configuration in configtx.yaml.

```
1  Channel1:
2    Consortium: SampleConsortium
3    Application:
4      <<: *ApplicationDefaults
5      Organizations:
6        - *Org1
7        - *Org2
8      Capabilities:
9        <<: *ApplicationCapabilities
```

FIGURE 12.26   Channel profile configuration in Configtx.yaml.

| Configtx tool binary | Name of the profile object of the orderer in the configtx.yaml | Tells the tool where to specifically output the block |

`../bin/configtxgen -profile OrdererGenesis -outputBlock ./channel-artifacts/genesis.block`

FIGURE 12.27   Command for genesis block.

Configtx tool binary
Name of the profile object for the
channel in the configtx.yaml

../bin/configtxgen -profile ${CHANNEL_PROFILE}

Tell the tool to specifically output the channel trasaction file to

-outputCreateChannelTx ./channel-artifacts/${CHANNEL_NAME}.tx

give the channel name for the channel transaction

-channelID $CHANNEL_NAME

FIGURE 12.28   Creating channel transaction file.

Figure 12.29 shows command which is used to generate anchor peer transaction files. When we created the channel transaction file, there is no anchor peer defined by default in the file. We will create a transaction file update for the channel. This transaction file will set the address and MSP of the anchor peer. We will use this file to update the channel configuration after creating the channel.

For creating the channel using org1 peer, we need to create channel using any one of the organizations and then join the channel using the peers of authorized organization.

Figure 12.30 shows command to create a channel. To create a channel, we will need the channel transaction file that is generated using the previous command. When we create the channel, it will generate the channel block, we will copy this channel block to the host network, the location which is persistent volume for all the peers, i.e., the channel block will be copied to all the peers, which is necessary for any block to join the channel.

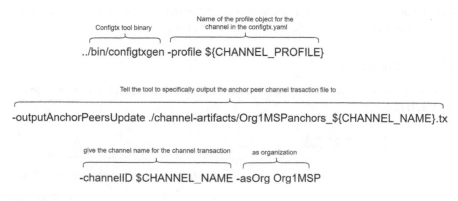

Configtx tool binary
Name of the profile object for the
channel in the configtx.yaml

../bin/configtxgen -profile ${CHANNEL_PROFILE}

Tell the tool to specifically output the anchor peer channel trasaction file to

-outputAnchorPeersUpdate ./channel-artifacts/Org1MSPanchors_${CHANNEL_NAME}.tx

give the channel name for the channel transaction   as organization

-channelID $CHANNEL_NAME -asOrg Org1MSP

FIGURE 12.29   Command for anchor peers.

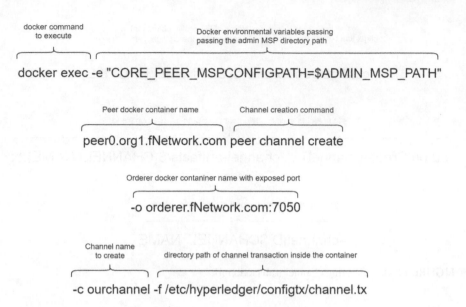

FIGURE 12.30   Creating channel.

After executing the above command shown in Figure 12.31, each and every peer will have the channel block inside their container file system due to the concept of docker volumes.

Now using the channel block, a peer can join a channel. The below command is used for a peer to join the channel

Now we have successfully joined using command shown in Figure 12.32, and the channel using peer0 of org1; now repeat the same for each peer of all organizations which are in the channel consortium by changing the peer docker container name.

After joining the channel using the command shown in Figure 12.33 successfully, we need to update the anchor peer details for the organization. For this, we will use the anchor peer transaction update file generated using the previous command. As we have only peer for each organization, the same peer acts as the anchor peer. Use the above command template and change the peer docker container name and anchor channel transaction file path for each organization.

12.4.6   CHAINCODE

In this subsection, we will discuss about the development of chaincode (smart contract) in NodeJS.

docker exec peer0.org1.fNetwork.com cp ourchannel.block /etc/hyperledger/configtx

FIGURE 12.31   Display command.

FIGURE 12.32 Command to join a peer.

FIGURE 12.33 Anchor peers update command.

Make a new directory outside of Fabric, name it chaincode. Open the terminal and go to the chaincode folder path. Run "npm init, "this command initializes the NodeJS project. After entering the details of npm init, run "npm i fabric-shim."

Fabric-shim: The fabric-shim provides the framework for the chaincode, a lower-level API for "smart contracts" implementation.

By using fabric-shim NodeJS module, we will develop a chaincode, which transfers an asset from one organization to another with a message attached to it.

Some of the objects and functions by fabric-shim are as follows.

12.4.6.1   Stub, Getstate, Putstate, getQueryResult, getHistoryForKey

Create a file index.js and write the following code in it.

```
1. 'use strict';
2.
3. const shim = require('fabric-shim');
4. const util = require('util');
5.
6. async function queryByKey(stub, key) {
7.    console.log('=========== START : queryByKey ===========');
8.    console.log('##### queryByKey key: ' + key);
9.
10. let resultAsBytes = await stub.getState(key);
11. if (!resultAsBytes || resultAsBytes.toString().length <= 0) {
12.    throw new Error('##### queryByKey key: ' + key + ' does
       not exist');
13. }
14. console.log('##### queryByKey response: ' +
    resultAsBytes);
15. console.log('=========== END : queryByKey ===========');
16. return resultAsBytes;
17. }
18.
19. let Chaincode = class {
20.   async Init(stub) {
21.   console.log(
22.   '=========== Init: Instantiated / Upgraded ngo chaincode
       ==========='
23. );
24.   return shim.success();
25. }
26.   async Invoke(stub) {
27.   console.log('=========== START : Invoke ===========');
28.   let ret = stub.getFunctionAndParameters();
29.   console.log('##### Invoke args: ' + JScON.stringify(ret));
30.
31.   let method = this[ret.fcn];
32.   if (!method) {
33.    console.error(
34.    '##### Invoke - error: no chaincode function with name:
       ' +
35.    ret.fcn +
36.    ' found'
37. );
38.     throw new Error('No chaincode function with name: ' +
       ret.fcn + ' found');
39.   }
40.   try {
41.    let response = await method(stub, ret.params);
42.    console.log('##### Invoke response payload: ' +
       response);
43. return shim.success(response);
```

```
44.     } catch (err) {
45.     console.log('##### Invoke - error: ' + err.stack);
46.     return shim.error(err);
47.     }
48. }
49.   async initLedger(stub, args) {
50.     console.log('============ START : Initialize Ledger
        ===========');
51.     console.log('============ END : Initialize Ledger
        ===========');
52. }
53.     async createAsset(stub, args) {
54.       try{
55.         args = JSON.parse(args);
56.       let asset = {
57.       docType: "project",
58.       assetOwner: args.owner,
59.       transferMessage: ""
60.       }
61.       let resultAsBytes = await stub.getState(args.assetID.
        toString());
62.       if (!(!resultAsBytes || resultAsBytes.toString().length
        <= 0)) {
63.       throw new Error('##### asset with assetID: ' + key + '
        exists');
64.       }
65.       const buffer = Buffer.from(JSON.stringify(asset));
66.       await stub.putState(args.assetID.toString(), buffer);
67.       }catch(e){
68.         console.log(e);
69.         return shim.error(e);
70.     }
71. }
72.
73.     async getAsset(stub, args){
74.       try{
75.         args = JSON.parse(args);
76.         return await queryByKey(stub, args.assetID);
77.         }catch(e){
78.       console.log(e);
79.       return shim.error(e);
80.     }
81. }
82.
83.   async transferAsset(stub, args){
84.     try{
85.     args = JSON.parse(args);
86.     let asset = await queryByKey(stub, args.assetID);
87.     asset = JSON.parse(asset);
88.     if(asset.assetOwner!== args.userName){
89.     throw new Error("asset is not owned by user who is
        executing this transaction");
```

```
90.  } else {
91.    asset.assetOwner = args.transferTo;
92.    asset.transferMessage = args.transferMessage;
93.    const buffer = Buffer.from(JSON.stringify(asset));
94.    await stub.putState(args.assetID.toString(), buffer);
95.  }
96.    }catch(e){
97.
98.  }
99.  }
100.
101.  async deleteAsset(stub, args) {
102.    args = JSON.parse(args);
103.    await stub.deleteState(args.assetID);
104. }
105.};
106.  shim.start(new Chaincode()); Using the above chaincode
       we will install the chaincode on the peers and
       instantiate the chaincode.
```

The Figure 12.34 displays command for installing a chaincode.
The Figure 12.35 displays command for instantiating a chaincode.

12.4.7   REST API

In this section, we are going to discuss about Fabric SDK and how to create REST API to interact with the blockchain network using chaincode.

12.4.7.1   Fabric SDK

Hyperledger Fabric aims to currently provide SDKs for two languages Node and Java.

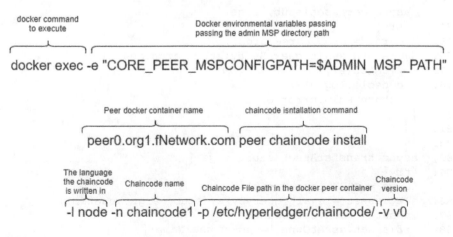

FIGURE 12.34   Install chaincode command.

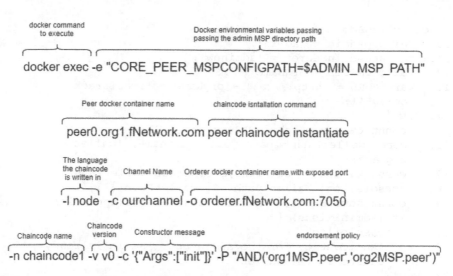

docker command
to execute

Docker environmental variables passing
passing the admin MSP directory path

docker exec -e "CORE_PEER_MSPCONFIGPATH=$ADMIN_MSP_PATH"

Peer docker container name   chaincode isntallation command

peer0.org1.fNetwork.com peer chaincode instantiate

The language
the chaincode   Channel Name   Orderer docker contaniner name with exposed port
is written in

-l node   -c ourchannel -o orderer.fNetwork.com:7050

                Chaincode
Chaincode name version   Constructor message   endorsement policy

-n chaincode1 -v v0 -c '{"Args":["init"]}' -P "AND('org1MSP.peer','org2MSP.peer')"

FIGURE 12.35 Instantiate chaincode command.

In this walkthrough, we are using the Hyperledger Fabric SDK for NodeJS. This SDK provides a means to communicate with the blockchain, interact with smart contracts, i.e., chaincode deployed on the Fabric network.

Fabric-network: It provides user applications with high-level APIs to send transactions and test smart contract queries (chaincode).

API: Application developer pluggable APIs for providing alternative implementations of key SDK interfaces. Default implementations are included for each interface.

Fabric-client: It offers APIs to communicate with Hyperledger's core blockchain network elements, namely peers, orderers, and event streams.

Fabric-ca-client: It provides APIs to interact with the fabric-ca optional component, which includes membership management services.

12.4.7.2   User Interaction API using the Fabric SDK (NodeJS)

In this section, we will discuss about four major operations/functions:

1. Register – registering a client.
2. Query – querying blockchain.
3. Invoke – invoking blockchain.
4. Block Listener–listen to a generated block.

12.4.7.2.1   Registering a Client

```
1. const FabricCAServices = require('fabric-ca-client');
2. const path = require('path');
3. const {
4. FileSystemWallet,
5. Gateway,
6. X509WalletMixin
7. } = require('fabric-network');
8. const fs = require('fs');
```

```
9. const regAdmin = async orgName => {
10.  var appAdmin = 'admin';
11.  var appAdminSecret = 'adminpw';
12.  var orgMSPID = orgName + 'MSP';
13.  var caURL = 'http://ca.' + orgName + '.meditrack.
     com:7054';
14. try {
15.    const ca = new FabricCAServices(caURL);
16.    const walletPath = path.join(__dirname, 'wallet',
       orgName);
17.    const wallet = new FileSystemWallet(walletPath);
18.    console.log(`Wallet path: ${ walletPath }`);
19.    const adminExists = await wallet.exists(appAdmin);
20.    if (adminExists) {
21.      console.log(
22.        'An identity for the admin user "admin" already
           exists in the wallet'
23.      );
24.    return 'already admin exists';
25.    }
26.    const enrollment = await ca.enroll({
27.      enrollmentID: appAdmin,
28.      enrollmentSecret: appAdminSecret
29.    });
30.    const identity = X509WalletMixin.createIdentity(
31.      orgMSPID,
32.      enrollment.certificate,
33.      enrollment.key.toBytes()
34.    );
35.    wallet.import(appAdmin, identity);
36.    console.log(
37.      'msg: Successfully enrolled admin user ' +
38.      appAdmin +
39.      'and imported it into the wallet'
40.    );
41.    } catch (e) {
42.      console.log(e.stack);
43.    }
44. };
45. const regUser = async (orgName, userName) => {
46.    const ccpPath = path.join(__dirname, '..', 'config',
       orgName, 'connectionProfile.json');
47.    const ccpJSONContenet = fs.readFileSync(ccpPath,
       'utf8');
48.    const ccp = JSON.parse(ccpJSONContenet);
49.    try {
50.    let appAdmin = 'admin';
51.    let orgMSPID = orgName + 'MSP';
52.    let gatewayDiscovery = {
53.      enabled: true,
54.      asLocalhost: false
```

```
55.    };
56.    const walletPath = path.join(__dirname, 'wallet',
       orgName);
57.    const wallet = new FileSystemWallet(walletPath);
58.    console.log(`Wallet path: ${ walletPath }`);
59.    const userExists = await wallet.exists(userName);
60.    if (userExists) {
61.      console.logimport(
62.      `An identity for the user ${ userName } already exists
         in the wallet`
63.    );
64.    return 'User already registered';
65.    }
66.    const adminExists = await wallet.exists(appAdmin);
67.    if (!adminExists) {
68.      console.log(
69.        `An identity for the admin user ${ appAdmin } does not
           exist in the wallet`
70.    );
71.      console.log('Registering the admin');
72.      regAdmin(orgName);
73.    }
74.    const gateway = new Gateway();
75.    await gateway.connect(ccp, {
76.      wallet,
77.      identity: appAdmin,
78.      discovery: gatewayDiscovery
79.    });
80.    const ca = gateway.getClient().
       getCertificateAuthority();
81.    const adminIdentity = gateway.getCurrentIdentity();
82.    const secret = await ca.register(
83.    {
84.    affiliation: 'org1',
85.    enrollmentID: userName,
86.    role: 'client'
87.    },
88.    adminIdentity
89.    );
90.
91.    console.log('secret:');
92.    console.log(secret);
93.    const enrollment = await ca.enroll({
94.    enrollmentID: userName,
95.      enrollmentSecret: secret
96.    });
97.    const userIdentity = X509WalletMixin.createIdentity(
98.      orgMSPID,
99.      enrollment.certificate,
100.     enrollment.key.toBytes()
101.   );
```

```
102.    wallet.import(userName, userIdentity);
103.    console.log(
104.    'Successfully registered and enrolled admin user ' +
105.    userName +
106.    ' and imported it into the wallet'
107.    );userName
108.    return {
109.       userName,
110.       secret
111.    }
112.    } catch (e) {
113.      console.log('Error Registering the user' + e.stack);
114.      return e
115.    }
116.    };
117.    // regUser('bayer', 'user2');
118.
119.    module.exports = regUser;
```

The above function regUser uses several API's provided by Fabric SDK. It begins with wallet, as wallet stores the private key to sign transactions. The wallet is obtained and then the system verifies if a user already exists. If a user already exists, then it returns a message that user already exists. If not, then a user is registered and respective certificates are released and added to the wallet. If an organization doesn't have an admin and user is the first to register, then user is registered as admin first and then certificates are generated.

12.4.7.2.2 Querying

```
1. 'use strict';
2. const { FileSystemWallet, Gateway } =
   require('fabric-network');
3. const fs = require('fs');
4. const path = require('path');
5.
6. async function query(functionName, args, channelName,
   contractName, orgName, userName) {
7.  let gateway
8.  try {
9.    const ccpPath = path.join(
10.     __dirname,
11.     '..',
12.     'config',
13.     orgName,
14.     'connectionProfile.json'
15.   );
16.    const ccpJSONContent = fs.readFileSync(ccpPath, 'utf8');
17.    const ccp = JSON.parse(ccpJSONContent.toString());
18.
19.    const walletPath = path.join(__dirname, 'wallet',
       orgName);
```

```
20.    const wallet = new FileSystemWallet(walletPath);
21.    console.log(`Wallet path: ${ walletPath }`);
22.
23.     const userExists = await wallet.exists(userName);
24.    if (!userExists) {
25.     console.log(
26.     'An identity for the user ' + userName + ' does not
        exist in the wallet'
27.    );
28.     console.log('Run the registerUser.js application before
        retrying');
29. throw new Error('An identity for the user ' + userName + '
    does not exist in the wallet');
30.    }
31.
32.    gateway = new Gateway();
33.    await gateway.connect(ccp, {
34.     wallet,
35.     identity: userName,
36.     discovery: { enabled: false }
37.    });
38.    const network = await gateway.getNetwork(channelName);
39.
40.    const contract = network.getContract(contractName);
41.    args = JSON.stringify(args);
42.    let result = await contract.evaluateTransaction(function
       Name, args);
43.
44.    console.log(
45.    `Transaction has been evaluated, result is: ${ result.
       toString() }`
46.    );
47.    result = result.toString();
48.    return result;
49. } catch (error) {
50.    console.error(`Failed to evaluate transaction: ${ error
       }`);
51.    return error;
52.    } finally {
53.    if (gateway) {
54.     await gateway.disconnect();
55. }
56. }
57. }
58.
59. module.exports = query;
```

We can query the blockchain ledger using the above function. This function uses
several API's from Fabric SDK and gives the result of the query

12.4.7.2.3   Invoking a Transaction

```
1.   const path = require('path');
2.   const { FileSystemWallet, Gateway } =
     require('fabric-network');
3.   const fs = require('fs');
4.   const util = require('util');
5.
6.   async function invoke(functionName, args, channelName,
     contractName, orgName, userName) {
7.    let demo_gateway;
8.    try {
9.     const ccpPath = path.join(
10.       __dirname,
11.      '..',
12.      'config',
13.      orgName,
14.      'connectionProfile.json'
15.    );
16.     const ccpJSONContents = fs.readFileSync(ccpPath,
        'utf8');
17.     const ccp = JSON.parse(ccpJSONContents.toString());
18.
19.     const walletPath = path.join(__dirname, 'wallet',
        orgName);
20.     const wallet = new FileSystemWallet(walletPath);
21.     console.log(`Wallet path: ${ walletPath }`);
22.
23.     const userExists = await wallet.exists(userName);
24.     if (!userExists) {
25.        console.log(
26. '     An identity for the user ' + userName + ' does not
           exist in the wallet'
27. );
28.        console.log('Run the registerUser.js application
           before retrying');
29. throw new Error('An identity for the user ' + userName + '
     does not exist in the wallet');
30.    }
31.
32.       demo_gateway = new Gateway();
33.       await demo_gateway.connect(ccp, {
34.         wallet,
35.         dentity: userName,
36.         discovery: { enabled: false }
37. });
38.        const network = await demo_gateway.
           getNetwork(channelName);
39.
```

```
40.     const contract = network.getContract(contractName);
41.
42.      let txId, status, blockNo;
43.
44.     let tx = await contract.createTransaction(functionN
        ame);
45.     let proResolve;
46.     let pro = new Promise((resolve, reject) => {
47.     proResolve = resolve;
48.   });
49.     const lis = await tx.addCommitListener((error,
        transactionId, status, blockNumber) => {
50.     if (error) {
51.       console.error(error);
52.       proResolve();
53.       throw new Error(error);
54.     }
55.     txId = transactionId;
56.     status = status;
57.     blockNo = blockNumber;
58.     proResolve();
59.   })
60.   args = JSON.stringify(args);
61.   let result = await tx.submit(args);
62.   console.log(
63.     `Transaction has been submitted, result is: ${ result.
        toString() }`
64.   );
65.   await pro
66.   return {
67.     result: result.toString(),
68.     txId,
69.     status,
70.     blockNo
71.     }
72.   } catch (error) {
73.     console.error(`Failed to evaluate transaction: ${
        error }`);
74.     return error
75.     } finally {
76.     if (gateway) {
77.       await demo_gateway.disconnect();
78.     }-
79.   }
80. }
81. module.exports = invoke;
```

Using Fabric SDK, we can invoke a transaction and get the invoked transaction id.

These are some basic functions to get started with a demo API. Furthermore, using Fabric SDK custom API's can be created, depending on our use case.

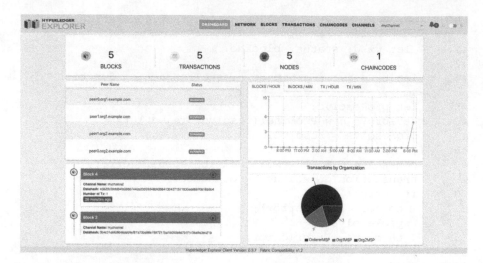

FIGURE 12.36 Explorer structure.

HYPERLEDGER EXPLORER   Select Channel ▾

BLOCKLIST ⌄ ↻ ✕		BLOCK #4	✎ ⌄ ↻ ✕
Block	**TXNs**	number	4
#4	1	previous_hash	968ccf2e06257259ca59b181b9196e360d42919da1fb57c44bc513b9785de0c6
#3	1	data_hash	3091a9fc3d4972981471d91e94eca77a1e057e1cdd2e0fa86bbd0f71d2284a07
#2	1	Transactions	590a31e86a673221706d46dd0f3a89ae8249abaa0889cf5a37066f0a65c4ed39
#1	1		
#0	1		

FIGURE 12.37 Block structure.

12.4.8 HYPERLEDGER EXPLORER

Hyperledger Explorer is a tool developed by the Linux foundation. Designed to view, invoke, deploy, or query blocks, transactions and associated data, network information (name, status, list of nodes), chaincodes and transaction families, as well as any other relevant information stored in the ledger.

In this project, we implemented the Hyperledger Explorer by using the docker image of the explorer.

Hyperledger Explorer docker repository: https://hub.docker.com/r/hyperledger/explorer/

Hyperledger Explorer PostgreSQL docker repository: https://hub.docker.com/r/hyperledger/explorer-db

To run the Hyperledger Fabric using docker, we first need to create a network configuration file which has the details about the Hyperledger Fabric network such as the peer address and channel details.

The Landing Page of the Hyperledger Fabric explorer page gives the details such as the number of blocks created in the channel, the number of transactions in the channel, the nodes participating in the channel, the number of chaincodes installed in the channel, and graphical representation of the transactions taking place.

The snapshot in Figure 12.36 displays the interface of Hyperledger Fabric explorer.

The snapshot in Figure 12.37 displays the interface of Hyperledger Fabric explorer, which shows block.

REFERENCES

1. Hyperledger Fabric: A Distributed Operating System for Permissioned Blockchains. Elli Androulaki Artem Barger Vita Bortnikov IBM Christian Cachin Konstantinos Christidis Angelo De Caro David Enyeart IBM Christopher Ferris Gennady Laventman Yacov Manevich IBM Srinivasan Muralidharan* State Street Corp. Chet Murthy* Binh Nguyen* State Street Corp. Manish Sethi Gari Singh Keith Smith Alessandro Sorniotti IBM Chrysoula Stathakopoulou Marko Vukolić Sharon Weed Cocco Jason Yellick.
2. Blockchain White Paper. China Academy of Information and Communication Technology Trusted Blockchain Initiatives, December, 2018 China Academy of Information and Communication Technology Trusted Blockchain Initiatives December, 2018.
3. J. P. Morgan. Quorum whitepaper. https://github.com/jpmorganchase/ quorum-docs, 2016.
4. Buterin, V. Ethereum White Paper: A next-generation smart contract and decentralized application platform, 2013.
5. Ethereum: A Secure Decentralised Generalised Transaction Ledger-Dr. Gavin Wood Founder, Ethereum & Ethcore Gavin@Ethcore.Io.
6. BLOCKCHAIN AND SMART CONTRACTS Technologies, research issues and applications-Julian Schütte, Gilbert Fridgen, Wolfgang Prinz, Thomas Rose, Nils Urbach, Thomas Hoeren, Nikolas Guggenberger, Christian Welzel, Steffen Holly, Axel Schulte, Philipp Sprenge, Christian Schwede, Birgit Weimert, Boris Otto, Mathias Dalheimer, Markus Wenzel, Michael Kreutzer, Michael Fritz, Ulrich Leiner, Alexander Nouak.
7. Blockchain Technology Overview—Dylan Yaga Peter Mell Nik Roby Karen Scarfone p. 29.
8. Blockchains Architecture, Design and Use Cases Prof. Sandip Chakraborty Prof. Praveen Jayachandran Department of Computer Science and Engineering Indian Institute of Technology, Kharagpur Lecture—46 Blockchain Security—III (Fabric SideDB).

13 Fraud-Resistant Crowdfunding System Using Ethereum Blockchain

Sandeep Kumar Panda
ICFAI Foundation for Higher Education
(Deemed to be University)

CONTENTS

13.1 INTRODUCTION

Crowdfunding [1,2] is the practice of funding a project or venture by raising small amounts of money from a large number of people, typically via the Internet. Crowdfunding is a form of crowdsourcing and alternative finance. In 2015, a world-wide estimate totaling over US$134 billion was raised by crowdfunding. Although similar concepts can also be executed through mail-order subscriptions, benefit events, and other methods, the term crowdfunding refers to Internet-mediated registries. This modern crowdfunding model is generally based on three types of actors: the project initiator who proposes the idea or project to be funded, individuals, or groups who support the idea, and a moderating organization (the "platform") that brings the parties together to launch the idea, which is raising small amount of funds for a project from large people via Internet. It is an alternative form of finance. It supports entrepreneurship and bring the innovative ideas of people into life [2]. Although similar concepts can also be executed through mail-order subscriptions, benefit events, and other methods, the term crowdfunding refers to Internet-mediated registries. This modern crowdfunding model is generally based on three types of actors [1]: the project initiator who proposes the idea or project to be funded, individuals or groups who support the idea, and a moderating organization (the "platform") that brings the parties together to launch the idea. There are so many crowdfunding platforms in which Kickstarter [3, 4] is one of the most famous and widely used platforms. The Kickstarter is taken as a case study and explains its functioning and issues with it.

Crowdfunding [5–7] has been used to fund a wide range of for-profit, entrepreneurial ventures such as artistic and creative projects, medical expenses, travel, or community-oriented social entrepreneurship projects. "Kickstarter" [3, 4] is one of the famous crowdfunding platforms, which was taken as a case study and explained its working, problems and issues and how the blockchain technology solves these issues.

13.2 KICKSTARTER

Kickstarter [3] is an American public-benefit corporation based in Brooklyn, New York, that maintains a global crowdfunding platform focused on creativity and merchandising. The company's stated mission is to "help bring creative projects to life." Kickstarter has reportedly received more than $1.9 billion in pledges from 9.4 million backers to fund 257,000 creative projects, such as films, music, stage shows, comics, journalism, video games, technology, and food-related projects. People who back Kickstarter projects are offered tangible rewards or experiences in exchange for their pledges. This model traces its roots to subscription model of arts patronage, where artists would go directly to their audiences to fund their work. The main mission of Kickstarter community is to help bring creative projects to life. They measure their success as a company by how well they achieve that mission, not by the size of their profits. The report further focuses on the Kickstarter platform in the Literature survey. Then the report elaborates on the implementation of the crowdfunding with the use of blockchain platform Ethereum.

Kickstarter is a crowdfunding platform that allows users to raise money for a project and typically provides rewards for the backers. For example, if someone posts an idea of building a smart-watch with its description, usage, benefits, cost, the time for making it and rewards that a backer gets based on the contribution is specified. The people who likes the idea will fund the project by contributing some amount of money. When sufficient funding is reached, the person will start making the project and deliver it within the time he specified. After successful completion of product, the rewards will be sent to the people who supported to the project based on their contribution.

A Kickstarter [3, 8] project page should have a video and description that clearly explains what are they building, the motivation behind making it, why are they creating it, the minimum amount of money required for producing the product, the deadline for funding, and the rewards that a contributor will receive when the product is completed. The progress of manufacturing the product should be constantly updated in the project page. If the funding did not reach the sufficient amount by the deadline, no funds are collected. Only when the funding goal is reached the backers credit card will be charged.

Kickstarter charges a fee of 5% of the total funds raised on a project. And also, there is an additional fee of 3%–5% will be debited to process the payments. Kickstarter claims no proprietorship or control on the projects and the work generated by the project creators. The project pages on the website are stored forever and are available for the public. After sufficient funding is achieved, projects and uploaded content cannot be modified or deleted from the site. In the following section, a flow chart of the working functionality of Kickstarter platform is given, refer Figure 13.1.

13.3 GUIDELINES IN KICKSTARTER

To maintain its focus as a funding platform for creative projects, Kickstarter has outlined three guidelines for all project creators to follow: creators can fund projects only; projects must fit within one of the site's 113 creative categories; and creators must abide by the site's prohibited uses, which include charity and awareness campaigns [3, 8]. Kickstarter has additional requirements for hardware and product design projects. These include:

- Banning the use of photorealistic renderings and simulations demonstrating a product.
- Banning projects for genetically modified organisms.
- Limiting awards to single items or a "sensible set" of items relevant to the project (e.g., multiple light bulbs for a house).
- Requiring a physical prototype.
- Requiring a manufacturing plan.

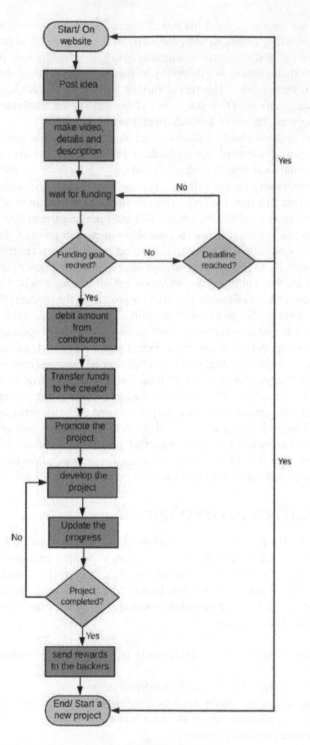

FIGURE 13.1 Flowchart of a project/campaign in Kickstarter.

The guidelines are designed to reinforce Kickstarter's position that people are backing projects, not placing orders for a product. To underscore the notion that Kickstarter is a place in which creators and audiences make things together, creators across all categories are asked to describe the risks and challenges a project faces in producing it. This educates the public about the project goals and encourages contributions to the community.

The Flow of a Project/Campaign on Kickstart

13.4 ISSUES WITH KICKSTARTER

There are many successful projects as well as many failure and fraudulent projects. There is no guarantee that people who post projects on Kickstarter will deliver on their projects, use the money to implement their projects, or that the completed projects will meet backers' expectations [1, 8]. Kickstarter advises backers to use their own judgment on supporting a project. They also warn project leaders that they could be liable for legal damages from backers for failure to deliver on promises. Projects might also fail even after a successful fundraising campaign when creators underestimate the total costs required or technical difficulties to be overcome. The backers will not get the refund for the failure product since the money is spent and the backers really cannot know whether it is a fraudulent project or it is a genuinely failed project. Either way the backers will not get any money return once they have contributed. All they have to do is wait for the reward or incentives that the project creator has promised. But there is a high percentage of possibility that even if the product is successfully completed, the backers may not receive the promised incentive. In those cases, Kickstarter won't do any help to the backers. If the backers want the product or the money they contributed, all they can do is file a law suit against the project creator which is done rarely since it also costs a lot of money.

There is always a possibility that a backer will always loose the money and not get justified in Kickstarter. There are so many obvious scams, fraud, and deceit in the Kickstarter [2, 4, 8]. Some of the problems with examples are illustrated below.

Security issues: There are some issues related to securely transferring the money to the project creator on the platform. This issue is explained by taking the case of a project known as "Pillow Talk" by Little Riot company [2, 8].

- Montgomery, the founder of Little Riot closed a successful Kickstarter campaign, after raising £82,000 in pre-orders for her invention Pillow Talk [9–11], which lets two people experience each other's "presence" by sending the sound of one person's real-time heartbeat to a speaker under their partner's pillow.
- The campaign was not only successful, it overfunded by almost 10%,but a glitch in Kickstarter's payments system meant that when it came to charging people's cards, every one of Pillow Talk's backers received an email saying their payment had been declined.

Emails came flooding in from people panicking that their payments hadn't processed. Montgomery says, "People were cancelling their debit cards and chaptering fraud to their banks. Every customer of one bank had their payments to us frozen."

- When Montgomery reported the issue to Kickstarter, the crowdfunding platform re-processed the payments but only 60% of the pledges came in, she claims. "People had deleted their details from the system. Kickstarter told us to chase those payments manually and that we had seven days to fix the pledges.
- According to Little Riot, Pillow Talk lost £20,000 of its total pledges as a result of the glitch but Kickstarter refused to waive its 5% commission.
- Joanna Montgomery posted in twitter saying, "I love it when you work for 5 years to launch your product and then @kickstarter fails to successfully process everyone's payments."

Copyright issues: Individuals steal their ideas from others. Therefore, trust on existing projects will be reduced. Monitoring this technically gets trust on the website. Constant monitoring of who is looking at your data and for how long is needed. Also, there are many patent issues in Kickstarter projects which lead to controversies and cancellation of projects [4, 8].

- On November 21, 2012, 13D Systems filed a patent infringement lawsuit against Formlabs and Kickstarter for infringing its 13D printer patent US 5,597,520, "Simultaneous multiple-layer curing in stereolithography."

Formlabs had raised $2.9 million in a Kickstarter campaign to fund its own competitive printer. The company said that Kickstarter caused "irreparable injury and damage" to its business by promoting the Form 1 printer, and taking a 5% cut of pledged funds. A six-month stay was granted by the judge for settlement talks in which Kickstarter did not participate.

- On September 13, 2011, Kickstarter filed a declaratory judgment suit against ArtistShare in an attempt to invalidate U.S. crowdfunding patent US 7885887 [12–15], "Methods and apparatuses for financing and marketing a creative work." Kickstarter asked that the patent be invalidated, or at the very least, that the court find that Kickstarter is not liable for infringement. In February 2012, ArtistShare and Fan Funded responded to Kickstarter's complaint by filing a motion to dismiss the lawsuit. They asserted that patent infringement litigation was never threatened, that "ArtistShare merely approached Kickstarter about licensing their platform, including patent rights," and that "rather than responding to ArtistShare's request for a counter-proposal, Kickstarter filed this lawsuit." The judge ruled that the case could go forward. ArtistShare then responded by filing a counterclaim alleging that Kickstarter was indeed infringing its patent. In June 2015, Kickstarter won its lawsuit with the judge declaring ArtistShare's patent invalid.

Controversies: There are many projects that are caught in the controversies because of the content, copying of ideas from a source or from a rival crowdfunding platform, etc. Some of these are mentioned below [8].

- In May 2014, Kickstarter blocked fundraising for a TV film about late-term abortionist Kermit Gosnell. Producer Phelim McAleer claimed Kickstarter censored the project because of its graphic content and espousing a "liberal agenda." In June 2014 [16, 17], the project received approval for fundraising from rival site Indiegogo, raising more than $2.13 million.
- In April 2013, filmmaker Zach Braff used Kickstarter to fund his film "Wish I Was Here" and raised $2 million in three days, citing the success of Rob Thomas' Veronica Mars Kickstarter as his inspiration. Braff received criticism for using the site, saying his celebrity status would draw attention from other creatives who lack celebrity recognition, the same kind of criticism regarding big figures in the gaming industry using Kickstarter. (One example is Richard Garriott, who created a successful $1+ million Kickstarter despite his personal fortune). Kickstarter disputed these arguments by claiming, according to their metrics, big name projects attract new visitors, who, in turn, pledge to lesser known projects.

Like this, there are so many problems surrounding Kickstarter and other similar crowdfunding platforms. From 2013, many Kickstarter projects are under the suspicion of generating fake backers or contributors to deceive the people into believing that the projects were successful, and to circumvent potential resources of similar funds. Many of Kickstarter problems can be solved with use of blockchain.

13.5   CROWDFUNDING WITH ETHEREUM BLOCKCHAIN

The main important thing in a crowdfunding platform is the secure use of the fund that is raised. Through Ethereum [2, 12, 18, 19], the money that is raised through the funding can be maintained and spent securely. In Kickstarter, when the funding goal is reached, the total funding amount is transferred to the campaign/project creator. But in this project, the total funding money is maintained and controlled by the smart contract. Whenever a user creates a campaign, a new contract related to that campaign is created which stores the money in ether. This contract manages all the tasks related to that campaign. This contract internally implements the voting system. In this project, assume there are only three kinds of people on the network, the manager who creates or hosts the projects, the contributors who invest the money in the project, and the vendors who sell items required for building a project. In this way, we can ensure the appropriate and efficient spending of the money that is raised through funding. The implementation of this is explained and demonstrated in further sections briefly.

13.6   WORKING OF CONTRACT

There are two contracts, namely CampaignFactory and Campaign. The Campaign Factory creates and manages all other contracts. The Campaign is an individual contract created specifically for a campaign/project. When a user hosts a campaign, a new campaign contract is created with some input variable like minimum contribution amount, the project manager address, and description about the project through CampaignFactory [1, 20] contract. The created contract accepts ether from the users and marks them as contributors. And the contributors will transfer ether to contract. When the manager wants to buy an item, he makes a request about the item by specifying the details like why the item is needed, the cost of the item, and the vendor address from where he wants to buy the item. Then the contributors will go for voting on this request, Figure 13.2 represents a brief model of this and if it gets greater than 51% of votes, the request will be approved and the ether will be transferred directly to the vendor and then the vendor will deliver the item to the manager. So, there is no need to be afraid of spending the money inappropriately or whether it would be taken away completely without doing anything. When making of the project completes, the manager will send the rewards to the contributors. Figure 13.2 shows the events and the flow of a campaign in the platform using Ethereum and smart contract (Figure 13.3).

13.6.1 Methods in CampaignFactory Contract

This contract is mainly used for creation and to keep track of the contracts. It consists of an array of type address to store the address of the deployed or created campaigns and two functions, one for the creation of the Campaign contract and the other for retrieving the addresses of the deployed campaigns. Figure 13.4 represents the structure of CampaignFactory contract.

Figure 13.5 shows the code involved in CampaignFactory contract. The method createCampaign has an argument minimum of type "uint." The "minimum" represents the minimum amount of ether required to contribute to become as a contributor for the campaign. "Public" is a keyword that indicates the access modifier or who can access this function. Inside this function, an address variable "newCampaign" is created for storing address value of the campaign deployed, which will be returned by the constructor of the contract Campaign. The constructor has two arguments "minimum" and "msg.sender." The variable "msg" is struct type which is automatically invoked when a transaction occurs or when a function is called and stores the details like the address of the person who invoked the transaction, the amount of ether he is sending with it, etc.

So, when a Campaign contract is being created, two variables representing the minimum contribution value and the address of the project creator or manager will be sent. The address returned which is stored in the newCampaign variable will be pushed into the array of deployed Campaigns variable.

In solidity, there are two types of functions:

1. The function that modifies the values in the contract or changes its state which is also known as transaction.
2. The function that does not change any value in the contract or changes its state which is also known as "call" to a function. To represent this type of function a keyword "view" or "pure" is attached in the function declaration.

The function "getDeployedCampaigns" is a view function that returns the addresses of the campaigns deployed on the Ethereum network, which are stored in the variable deployedCampaigns.

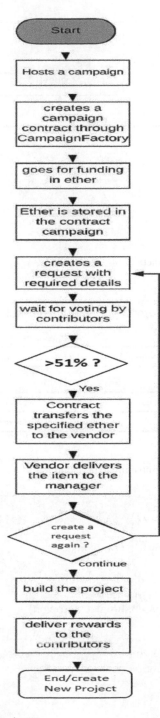

FIGURE 13.2 Flow of campaign.

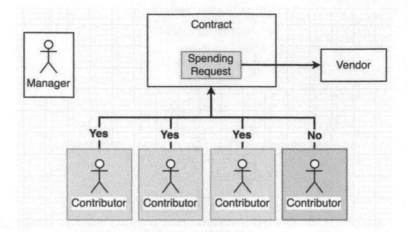

Yes

FIGURE 13.3   Model of contract.

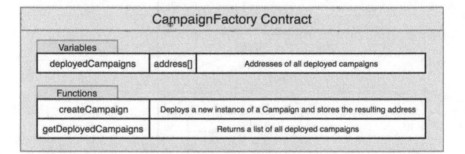

FIGURE 13.4   Structure of CampaignFactory contract.

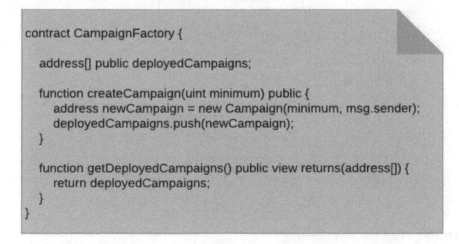

```
contract CampaignFactory {

    address[] public deployedCampaigns;

    function createCampaign(uint minimum) public {
        address newCampaign = new Campaign(minimum, msg.sender);
        deployedCampaigns.push(newCampaign);
    }

    function getDeployedCampaigns() public view returns(address[]) {
        return deployedCampaigns;
    }
}
```

FIGURE 13.5   CampaignFactory contract.

13.6.2   CONTRACT CAMPAIGN

Each contract created will represent a project on the front-end. This contract manages all the details and transferring of ether related to a project. This contract will be created by the CampaignFactory contract with two input arguments, minimum contribution value of type int and address of the manager of type address. The contract will store all the ether that is raised through the funding. In this, the participants will contribute some ether greater than the minimum value and become as contributors. The contributors have a right to vote for or against a request made by the manager. If the voting result is greater than 50%, then the request will be accepted and finally executed by the manager.

The Campaign contract has a struct of type Request which is shown in Figure 13.6. It contains the information about the requests that are made by the manager. The values for the variable's manager of type address and minimumContribution of type unit will be assigned through the constructor. The variable approvers is of type mapping between address and a Boolean value. It indicates whether an address is a contributor or not. The variable requests is an array of type Request struct which stores the list of requests created by the manager.

13.6.3   METHODS IN CAMPAIGN

Campaign (): This is a constructor function that will be invoked while creating the contract, and it has two arguments: minimum and creator. These two will be assigned to the variables minimumContribution and manager, refer Figure 13.7.

Modifier restricted (): Some functions like creating a request and finalizing request should only be done by the manager. So, these kinds of functions will have a similar code of checking whether the function is invoked by the manager or not. This verification can be kept in a modifier restricted function as shown in Figure 13.8. The restricted should be attached to the functions which will use this modifier.

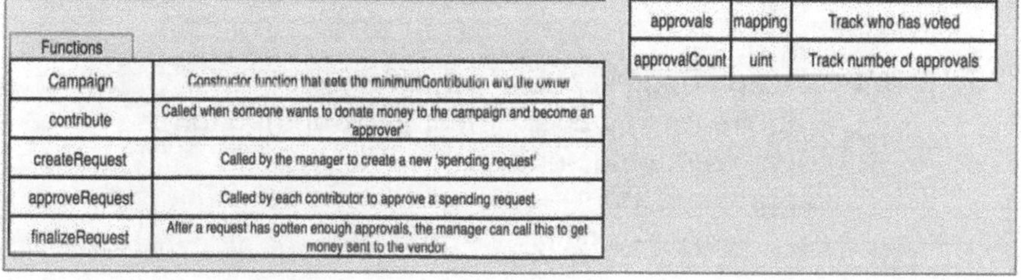

Campaign Contract

Variables				Request Struct		
manager	address	address of the person who is managing this campaign		description	string	Purpose of request
minimumContribution	uint	Minimum donation required to be considered a contributor or 'approver'		amount	uint	Ether to transfer
approvers	mapping	List of addresses for every person who has donated money		recipient	address	Who gets the money
requests	Request[]	List of requests that the manager has created.		complete	bool	Whether the request is done
				approvals	mapping	Track who has voted
Functions				approvalCount	uint	Track number of approvals
Campaign	Constructor function that sets the minimumContribution and the owner					
contribute	Called when someone wants to donate money to the campaign and become an 'approver'					
createRequest	Called by the manager to create a new 'spending request'					
approveRequest	Called by each contributor to approve a spending request					
finalizeRequest	After a request has gotten enough approvals, the manager can call this to get money sent to the vendor					

FIGURE 13.6   Structure of Campaign contract.

```
contract Campaign {

    struct Request {
        string description;
        uint value;
        address recipient;
        bool complete;
        uint approvalCount;
        mapping(address => bool) approvals;
    }

    Request[] public requests;
    address public manager;
    uint public minimumContribution;
    mapping(address => bool) public approvers;
    uint public approversCount;

    function Campaign(uint minimum, address creator) public {
        manager = creator;
        minimumContribution = minimum;
    }
//.............

}
```

FIGURE 13.7   Contract campaign code.

```
modifier restricted() {
    require(msg.sender == manager);
    _;
}
```

FIGURE 13.8   Modifier restricted.

The "require" will check whether the condition given is true or false. If it is true, the function will execute or else it will not be executed. The symbol "_,"represents that there is a condition to be executed.

Contribute (): This function has the access modifier public, which indicates that it can be accessed by anyone on the platform. The keyword "payable" is a modifier, which indicates that this function accepts ether, as shown in Figure 13.9. So, when

```
function contribute() public payable {
    require (msg.value > minimumContribution);
    approvers[msg.sender] = true;
    approversCount++;
}
```

FIGURE 13.9   Contribute method.

calling this function if the user has sent some ether, it will be directly added to the contract balance. Without the payable modifier, the function will not accept the ether that has been sent. The require will check the condition whether ether sent is greater than minimumContribution value. If it is then the function executes and he will be added as an approver or contributor. As mentioned above, approvers is a mapping between address and Boolean. The "msg.sender" will have the address of the person who invoked the function. For this address, the value true will be assigned which implies that he is a contributor. The variable approversCount is used to keep track of number of approvers are present for the project at a given time. Every time when some become an approver, the variable approversCount will be incremented by "1."

createRequest (): This function creates the request for spending the ether to buy something which will go for voting. The function should have the description or information about the request like the item name, why he want to buy it, the cost for the request, and the address of the recipient or vendor address to whose account the money should be deposited. This function has "restricted" attached to it as a modifier which indicates that the control will first go to the restricted modifier we have defined and execute it, refer Figure 13.10. So, it will verify whether the request is being created by the manager or not. If it is then it will execute the remaining function or else it won't.

Here, we are creating a variable "newRequest" of type Request struct which has a modifier "memory." The memory modifier indicates that this is a temporary variable and will be stored in the memory, only when the control is in the function or when the function is alive. The variables in the struct will be assigned with appropriate arguments given. Then this memory variable will be pushed into the requests array, which will be available for voting.

approveRequest (): The contributors or the approvers can approve a request by voting and if the approved percentage of a request is greater than 50%, the request will be accepted and can be finalized. The function has an argument index of type uint which represents the index of the request in requests array. In this method, a storage variable request of type Request struct is declared and is assigned with the value at the index of requests array. Storage means that this variable request points directly to the struct data at the index in the memory. It means if the data is changed

```
function createRequest(string description, uint value, address recipient)
    public restricted {

    Request memory newRequest = Request({   //Request(description, value, recipient, false)
        description: description,
        value: value,
        recipient: recipient,
        complete: false,
        approvalCount: 0
    });

    requests.push(newRequest);
}
```

FIGURE 13.10   createRequest method.

in the request variable, the data at the index of requests array will also be changed. In the next statement, we are verifying whether the function is invoked by one of the people (addresses) in the approvers mapping using require function. And also we are checking whether that person has already voted to the campaign by querying variable approvals in the Request struct that has the mapping between address and a Boolean type. If the person has already voted, the result will be true which will be false in the require condition and the function will not execute. Else, the approvals variable at the index will be assigned the value of true and approvalCount variable is incremented whenever a vote is accepted (Figure 13.11).

finalizeRequest (): The requests which got more than 50% approvals from the contributor can be executed by this function. This function will be executed, only when it is invoked by the manager of the project. The function is marked with the restricted modifier since only the manger should call the function. The function has an argument index of type uint which represents the request number. Again, a storage variable request of type Request struct is created and is assigned with the Request struct data at the index of the array requests, as shown in Figure 13.12. In the require statement, we are verifying whether the approvalCount is greater than 50%. We are also checking whether the request has already been completed to eliminate double spending. Finally, the amount in ether mentioned in the request will be transferred to the recipient address that is given while creating the request. Then the request will

```
function approveRequest(uint index) public {
    Request storage request = requests[index];
    require(approvers[msg.sender]);
    require(!requests[index].approvals[msg.sender]);
    request.approvals[msg.sender] = true;
    request.approvalCount++;
}
```

FIGURE 13.11   approveRequest method.

```
function finalizeRequest(uint index) public restricted {
    Request storage request = requests[index];
    require(request.approvalCount>(approversCount/2));
    require(!request.complete);
    request.recipient.transfer(request.value);
    request.complete = true;
}
```

FIGURE 13.12   finalizeRequest method.

be marked as complete by assigning the true value to the complete variable in the Request struct.

getSummary (): This function will return the summary or the current state of the contract.

This function is marked as view since it does not change the data in the contract. The functions return the minimum contribution value, balance of the contract or the fund of the contract, requests length, the number of contributors, and the manager of the contract.

Through the use of above-mentioned methods, the crowdfunding system spends the money efficiently and less chances for fraud can be achieved. In a brief, there are two contracts: one to control all the deployed campaigns contract and the other is for the campaign. The CampaignFactory contract will create contracts for the campaigns hosted on the site and stores the contract address. The contract campaign will control all the details and manages the money raised through funding for that particular campaign. The contract campaign has methods for contributing to the campaign through ether by becoming a contributor and raising funds, creating requests for purchasing thing from a vendor by the manager, voting for the request by the contributors, finalizing the request and transferring of ether, get the summary about the contract. All these methods will define the working of the contract. The sequence diagram for the whole process is given (refer Figure 13.13.)

The contract which we have designed should be deployed in the Ethereum network and a front-end must be created for user interaction purpose. For deployment, the test network like Rinkeby should be used first because it doesn't cost real ether.

13.7   CREATING THE FRONT-END USING REACT

As said in deployment section, after deploying the contract the address and the Application Binary Interface (ABI) of the contract will be stored in a file. The ABI and the address are used to call or invoke a function and sent a transaction. In the project, React [29] is used for the development of web page. For server and routing purposes, Next.js [28] is used. Semantic-ui-react [130] is used for the styling of the web page.

13.7.1   ARCHITECTURE OF NEXT.JS

Next.js is a module used for server and routing purposes [28]. Using the JavaScript will render in the server side and also on the client side, i.e., web browser. It wraps all the react and renders it in the browser. Figure 13.14 depicts the features and advantages of next.

When we use next, the default port of server is 13000. The next will render all the js files in the folder pages. When the server starts, the next will search for the index.js file in the pages folder. So, in the pages folder we are going to create all the required JavaScript files for the interaction with the contract. Figure 13.15 shows the brief model of how next routing works. The browser must have Metamask wallet installed, it is necessary for the website to work as it meant to be.

In the home page, we have the search option for campaigns and we will display all the campaigns that are deployed. A page will be created for hosting of new campaign

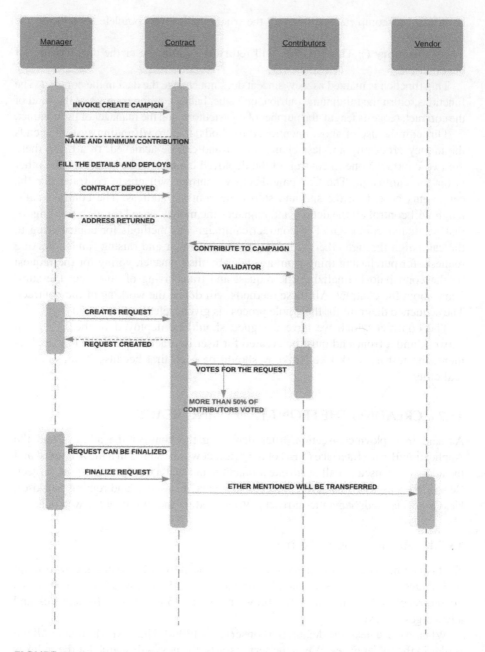

FIGURE 13.13   Sequence diagram.

and deploying the contract for that campaign. There will be a page or js file for details about the campaign and to contribute to it. For creating requests, viewing the requests, and finalizing requests, separate pages will be created for each of them. Figure 13.16 depicts the pages and the routes that are going to be created.

Next.js

Wraps up React + associated tools into one package

Lots of fancy features included out of the box	*Routing* *Server side rendering* *Hot module reload*

Makes it really, really easy to use React to make a multi-page application

FIGURE 13.14 Features of Next.js

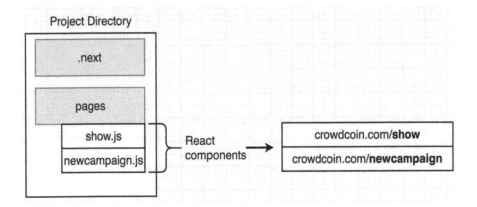

FIGURE 13.15 Next pages.

Routing	
Path	**We should show...**
/	List of Campaigns
/campaigns/new	Form to make a campaign
/campaigns/0x8147	Campaign details for campaign at address 0x8147
/campaigns/0x8147/requests	Requests for campaign at address 0x8147
/campaigns/0x8147/requests/new	Form to create a request for campaign at address 0x8147

FIGURE 13.16 Routing.

13.7.2 ROUTING AND FRONT-END

The index.js file in the pages folder will be the home page in which we will get the address of all the deployed campaigns through the CampaignFactory contract and will display all the campaigns. In this page, we also will keep a button for creating the new campaigns. Figure 13.17 shows the screenshot of the homepage.

The top menu bar will be there for all the pages as it is created as a header component and is imported to all the js files. The create campaign button will take you to the new campaign creation page where a form will be there. The web13 library will be used for the interaction with the contract along with the ABI. For displaying all the campaigns, we are accessing the getdeployedCampaigns method in the CampaignFactory contract through web13.

```
static async getInitialProps() {
    const campaigns = await factory.methods.getDeployedCampaigns().call();
    return { campaigns: campaigns };
}
```

While loading the page, we have to check whether there are any new campaigns deployed, for that purpose this function should be invoked every time when the page loads. The method "getInitialProps ()" will always be executed while loading the page in react. As the call to the method in the contract asynchronous async and await keywords are used. The "factory" is the interface to the methods in CampaignFactory contract. Using that getDeployedCampaigns () method has been called. The factory has been imported from the file where web13 has been interacting with the contract and it is stored in the instance which is exported

```
const instance = new web3.eth.Contract(
    JSON.parse(CampaignFactory.interface),
    '0x0A660371245d8Da49C525b54069A3B8E9F5031C5'
);
export default instance;
```

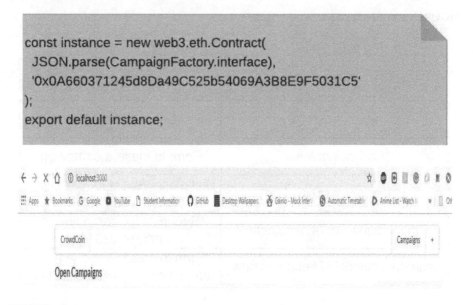

FIGURE 13.17   Homepage.

In the above code, the ABI and the address where the contract is deployed are directly pointed in the Ethereum network and are stored in the instance variable which is exported. The index.js imported this factory.js file and used this method while calling the getDeployedCampaigns () function.

```
const accounts = await web3.eth.getAccounts();
  await factory.methods.createCampaign(this.state.minimumContribution)
  .send({ from: accounts[0] });
```

For creating the new campaign, the code is written in the new.js file. A form will be there in the page where we have to fill all the details and mention the minimum contribution amount for the campaign. The web13 will take the first account present in the Metamask wallet and create the campaign and deploys the campaign under that address which will be the manager address for the campaign.

```
static async getInitialProps(props) {
      const campaign = Campaign(props.query.address);
      const summary = await campaign.methods.getSummary().call();
      return {
        address: props.query.address,
        minimumContribution: summary[0],
        balance: summary[1],
        requestsCount: summary[2],
        approversCount: summary[3],
        manager: summary[4]
      };
}
```

The details and mechanism for contribution of the contract will be created in the show.js file. When we click the "ViewCampaign" link (refer Figure 13.18) this file will be invoked, and the address is passed as an argument which will be used for retrieving the details of that campaign. As said above, the function mentioned in the above code will be executed whenever the page loads. The "props" is an argument that will be sent when this page is invoked which consists of the deployed contract address. Like factory campaign refers to the contract Campaign. Through this the method "getSummary ()" is called, and the current state of the campaign is retrieved. Using the information all details will be displayed.

FIGURE 13.18 Campaign page.

```
await campaign.methods.contribute().send({
    from: accounts[0],
    value:web3.utils.toWei(this.state.value, 'ether')
});
```

When the contribute button is clicked, the code referred here will be executed sending the amount of ether mentioned to the contract and becoming a contributor. The button "ViewRequests" will take to the page where all the requests created for that particular contract and their details and state will be displayed.

```
static async getInitialProps(props) {
    const { address } = props.query;
    const campaign = Campaign(address);
    const requestCount = await campaign.methods.getRequestsCount().call();
    const approversCount = await campaign.methods.approversCount().call();
    const requests = await Promise.all(
      Array(parseInt(requestCount)).fill().map((element, index) => {
        return campaign.methods.requests(index).call()
      })
    );

    return { address, requests, requestCount, approversCount };
}
```

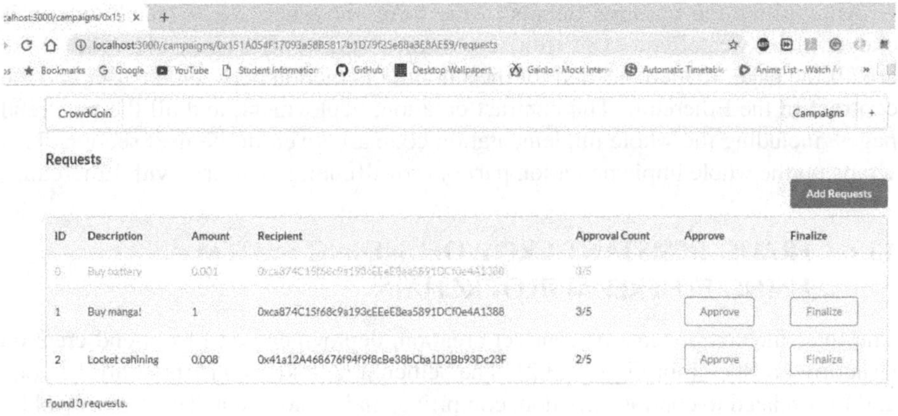

FIGURE 13.19 View requests page.

The above code will get all the requests and the details about the requests. This code will also be executed every time when the page loads and the props argument contains the address of the deployed campaign contract. The "getRequestsCount ()" will be called first getting the total requests count, and using this, a mapping of each request and the details is called through the Array and map function. The approval count for the requests will also be retrieved, and all these details will be displayed in the requests page. Figure 13.19 represents the screenshot of the request page.

All the request will be displayed in the table form and the requests which are fulfilled will be marked lighted and the requests which can be finalized will be shown in light gray color. The Approve and Finalize buttons will be present for each which corresponds to each request.

When Approve button is clicked, the "approveRequest ()" in the contract will be invoked, and likewise for the Finalize button which invokes the "FinalizeRequest ()" method. The button "AddRequests" will display a form for creating the new Request (refer Figure 13.20).

FIGURE 13.20 Create request page.

After filling the required details in the form and when the "Create!" button is clicked the "CreateRequest ()" function in the contract will be invoked. So, these are all the pages that are needed for the smooth interaction between the user and the contract in the Ethereum. The contract creation, deployment, and all the front-end pages, including the whole implementation code are given in the next section. This wraps up the whole implementation part of crowdfunding platform with Ethereum.

13.8 FRAUD-RESISTANT CROWDFUNDING SYSTEM USING ETHEREUM BLOCKCHAIN

The total file components of contract creation, deployment, and front-end creation structure are shown in Figure 13.21. The "ethereum" folder consists of all the code and files related to contract creation, compiling, and deployment. The "pages" folder consists of all front-end files and the components folder consists of component files which are used by the files in pages folder for the web page creation.

13.8.1 PACKAGE.JSON

This file shows all the information about the project, the installed dependencies, the start scripts, test scripts, and build scripts. This file gives an idea about the project and how to execute the project.

```
1   {
2       "name": "kickstart",
3       "version": "1.0.0",
4       "description": "",
5       "main": "index.js",
6       "scripts": {
7           "test": "mocha",
8           "dev": "node server.js",
9           "build": "next build"
10      },
11      "author": "",
12      "license": "ISC",
13      "dependencies": {
14          "fs-extra": "^5.0.0",
15          "ganache-cli": "^6.0.3",
16          "mocha": "^4.1.0",
17          "next": "^4.1.4",
18          "next-routes": "^1.2.0",
19          "react": "^16.2.0",
20          "react-dom": "^16.2.0",
21          "semantic-ui-css": "^2.2.12",
22          "semantic-ui-react": "^0.77.1",
23          "solc": "^0.4.19",
24          "truffle-hdwallet-provider": "0.0.3",
25          "web3": "1.0.0-beta.26"
26      }
27  }
```

FIGURE 13.21   Project file structure.

13.9   CONTRACT CODE, COMPILING, AND DEPLOYING

The contents and the file structure of the "ethereum" folder are shown in Figure 13.22.

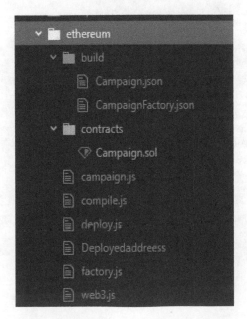

FIGURE 13.22   Ethereum file structure.

13.9.1 CAMPAIGN.SOL

```solidity
pragma solidity ^0.4.17;

contract CampaignFactory {
    address[] public deployedCampaigns;

    function createCampaign(uint minimum) public {
        address newCampaign = new Campaign(minimum, msg.sender);
        deployedCampaigns.push(newCampaign);
    }

    function getDeployedCampaigns() public view returns (address[]) {
        return deployedCampaigns;
    }
}

contract Campaign {
    struct Request {
        string description;
        uint value;
        address recipient;
        bool complete;
        uint approvalCount;
        mapping(address => bool) approvals;
    }
```

```solidity
    Request[] public requests;
    address public manager;
    uint public minimumContribution;
    mapping(address => bool) public approvers;
    uint public approversCount;

    modifier restricted() {
        require(msg.sender == manager);
        _;
    }

    function Campaign(uint minimum, address creator) public {
        manager = creator;
        minimumContribution = minimum;
    }

    function contribute() public payable {
        require(msg.value > minimumContribution);

        approvers[msg.sender] = true;
        approversCount++;
    }
```

```solidity
49        function createRequest(string description, uint value, address recipient) public restricted {
50            Request memory newRequest = Request({
51                description: description,
52                value: value,
53                recipient: recipient,
54                complete: false,
55                approvalCount: 0
56            });
57
58            requests.push(newRequest);
59        }
60
61        function approveRequest(uint index) public {
62            Request storage request = requests[index];
63
64            require(approvers[msg.sender]);
65            require(!request.approvals[msg.sender]);
66
67            request.approvals[msg.sender] = true;
68            request.approvalCount++;
69        }
70
71        function finalizeRequest(uint index) public restricted {
72            Request storage request = requests[index];
```

```solidity
73
74            require(request.approvalCount > (approversCount / 2));
75            require(!request.complete);
76
77            request.recipient.transfer(request.value);
78            request.complete = true;
79        }
80
81        function getSummary() public view returns (
82          uint, uint, uint, uint, address
83          ) {
84          return (
85            minimumContribution,
86            this.balance,
87            requests.length,
88            approversCount,
89            manager
90          );
91        }
92
93        function getRequestsCount() public view returns (uint) {
94            return requests.length;
95        }
96    }
```

13.9.2   COMPILE.JS

```
1   const path = require('path');
2   const solc = require('solc');
3   const fs = require('fs-extra');
4
5   const buildPath = path.resolve(__dirname, 'build');
6   fs.removeSync(buildPath);
7
8   const campaignPath = path.resolve(__dirname, 'contracts', 'Campaign.sol');
9   const source = fs.readFileSync(campaignPath, 'utf8');
10  const output = solc.compile(source, 1).contracts;
11
12  fs.ensureDirSync(buildPath);
13
14  for (let contract in output) {
15    fs.outputJsonSync(
16      path.resolve(buildPath, contract.replace(':', '') + '.json'),
17      output[contract]
18    );
19  }
```

13.9.3   DEPLOY.JS

```
1   const HDWalletProvider = require('truffle-hdwallet-provider');
2   const Web3 = require('web3');
3   const compiledFactory = require('./build/CampaignFactory.json');
4
5   const provider = new HDWalletProvider(
6     'call glow acoustic vintage front ring trade assist shuffle mimic volume reject',
7     'https://rinkeby.infura.io/orDImgKRzwNrVCDrAk5Q'
8   );
9   const web3 = new Web3(provider);
10
11  const deploy = async () => {
12    const accounts = await web3.eth.getAccounts();
13
14    console.log('Attempting to deploy from account', accounts[0]);
15
16    const result = await new web3.eth.Contract(
17      JSON.parse(compiledFactory.interface)
18    )
19      .deploy({ data: compiledFactory.bytecode })
20      .send({ gas: '1000000', from: accounts[0] });
21
22    console.log('Contract deployed to', result.options.address);
23  };
24  deploy();
```

13.9.4 WEB13.JS

```
1   import Web3 from 'web3';
2
3   let web3;
4
5   if (typeof window !== 'undefined' && typeof window.web3 !== 'undefined') {
6     // We are in the browser and metamask is running.
7     web3 = new Web3(window.web3.currentProvider);
8   } else {
9     // We are on the server *OR* the user is not running metamask
10    const provider = new Web3.providers.HttpProvider(
11      'https://rinkeby.infura.io/orDImgKRzwNrVCDrAk5Q'
12    );
13    web3 = new Web3(provider);
14  }
15
16  export default web3;
```

13.9.5 FACTORY.JS

```
1   import web3 from './web3';
2   import CampaignFactory from './build/CampaignFactory.json';
3
4   const instance = new web3.eth.Contract(
5     JSON.parse(CampaignFactory.interface),
6     '0xCA7740C40E82f945D4e48b9Cf2475c2674B2813D'
7   );
8
9   export default instance;
```

13.9.6 CAMPAIGN.JS

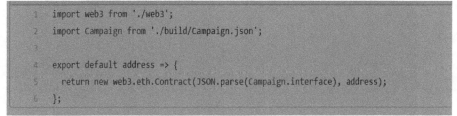

```
1   import web3 from './web3';
2   import Campaign from './build/Campaign.json';
3
4   export default address => {
5     return new web3.eth.Contract(JSON.parse(Campaign.interface), address);
6   };
```

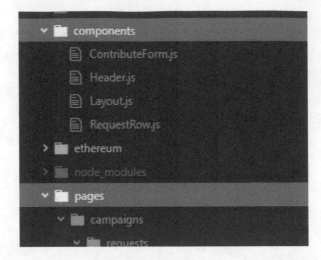

FIGURE 13.23 Front-end file structure.

13.10 FRONT END (REACTJS)

The "pages" and "components" folders consist of all the files for the creation of website. The components folder consists of component files which are required by the files in pages folder. This also requires a server.js and routes.js files for server and routing purposes. Figure 13.23 shows the file structures of the front-end.

13.10.1 SERVER.JS

```
1   const { createServer } = require('http');
2   const next = require('next');
3   const app = next({
4     dev: process.env.NODE_ENV !== 'production',
5     conf: {
6       webpack: config => {
7         config.devtool = false;
8         for (const r of config.module.rules) {
9           if (r.loader === 'babel-loader') {
10            r.options.sourceMaps = false;
11          } }
12        return config;
13      } }
14  });
15  const routes = require('./routes');
16  const handler = routes.getRequestHandler(app);
17  app.prepare().then(() => {
18    createServer(handler).listen(3000, err => {
19      if (err) throw err;
20      console.log('Ready on localhost:3000');
21    });
22  });
23
```

13.10.2 Routes.js

```
1    const routes = require('next-routes')();
2    routes
3      .add('/campaigns/new', '/campaigns/new')
4      .add('/campaigns/:address', '/campaigns/show')
5      .add('/campaigns/:address/requests', '/campaigns/requests/index')
6      .add('/campaigns/:address/requests/new', '/campaigns/requests/new');
7    module.exports = routes;
8
```

13.10.3 Components/Header.js

```
1    import React from 'react';
2    import { Menu } from 'semantic-ui-react';
3    import { Link } from '../routes';
4    export default () => {
5      return (
6        <Menu style={{ marginTop: '10px' }}>
7          <Link route="/">
8            <a className="item">CrowdCoin</a>
9          </Link>
10         <Menu.Menu position="right">
11           <Link route="/">
12             <a className="item">Campaigns</a>
13           </Link>
14           <Link route="/campaigns/new">
15             <a className="item">+</a>
16           </Link>
17         </Menu.Menu>
18       </Menu>
19     );
20   };
21
```

13.10.4 Components/Layout.js

```
1    import React from 'react';
2    import { Container } from 'semantic-ui-react';
3    import Head from 'next/head';
4    import Header from './Header';
5    export default props => {
6      return (
7        <Container>
8          <Head>
9            <link
10             rel="stylesheet"
11             href="//cdnjs.cloudflare.com/ajax/libs/semantic-ui/2.2.12/semantic.min.css"
12           />
13         </Head>
14         <Header />
15         {props.children}
16       </Container>
17     );
18   };
19
```

13.10.5   COMPONENTS/CONTRIBUTEFORM.JS

```
1   import React, { Component } from 'react';
2   import { Form, Input, Message, Button } from 'semantic-ui-react';
3   import Campaign from '../ethereum/campaign';
4   import web3 from '../ethereum/web3';
5   import { Router } from '../routes';
6   class ContributeForm extends Component {
7     state = {
8       value: '',
9       errorMessage: '',
10      loading: false
11    };
12    onSubmit = async event => {
13      event.preventDefault();
14      const campaign = Campaign(this.props.address);
15      this.setState({ loading: true, errorMessage: '' });
16      try {
17        const accounts = await web3.eth.getAccounts();
18        await campaign.methods.contribute().send({
19          from: accounts[0],
20          value: web3.utils.toWei(this.state.value, 'ether')
21        });
22        Router.replaceRoute(`/campaigns/${this.props.address}`);
23      } catch (err) {
24        this.setState({ errorMessage: err.message });
25      }
26      this.setState({ loading: false, value: '' });
```

```
27    };
28    render() {
29      return (
30        <Form onSubmit={this.onSubmit} error={!!this.state.errorMessage}>
31          <Form.Field>
32            <label>Amount to Contribute</label>
33            <Input
34              value={this.state.value}
35              onChange={event => this.setState({ value: event.target.value })}
36              label="ether"
37              labelPosition="right"
38            />
39          </Form.Field>
40          <Message error header="Oops!" content={this.state.errorMessage} />
41          <Button primary loading={this.state.loading}>
42            Contribute!
43          </Button>
44        </Form>
45      );
46    }}
47  export default ContributeForm;
48
```

13.10.6 COMPONENTS/REQUESTROW.JS

```
1   import React, { Component } from 'react';
2   import { Table, Button } from 'semantic-ui-react';
3   import web3 from '../ethereum/web3';
4   import Campaign from '../ethereum/campaign';
5   class RequestRow extends Component {
6     onApprove = async () => {
7       const campaign = Campaign(this.props.address);
8       const accounts = await web3.eth.getAccounts();
9       await campaign.methods.approveRequest(this.props.id).send({
10        from: accounts[0]
11      });
12    };
13    onFinalize = async () => {
14      const campaign = Campaign(this.props.address);
15      const accounts = await web3.eth.getAccounts();
16      await campaign.methods.finalizeRequest(this.props.id).send({
17        from: accounts[0]
18      });
19    };
20    render() {
21      const { Row, Cell } = Table;
22      const { id, request, approversCount } = this.props;
23      const readyToFinalize = request.approvalCount > approversCount / 2;
24      return (
25        <Row
```

```
26            disabled={request.complete}
27            positive={readyToFinalize && !request.complete}
28          >
29            <Cell>{id}</Cell>
30            <Cell>{request.description}</Cell>
31            <Cell>{web3.utils.fromWei(request.value, 'ether')}</Cell>
32            <Cell>{request.recipient}</Cell>
33            <Cell>
34              {request.approvalCount}/{approversCount}
35            </Cell>
36            <Cell>
37              {request.complete ? null : (
38                <Button color="green" basic onClick={this.onApprove}>
39                  Approve
40                </Button>
41              )}
42            </Cell>
43            <Cell>
44              {request.complete ? null : (
45                <Button color="teal" basic onClick={this.onFinalize}>
46                  Finalize
47                </Button>
48              )}
49            </Cell>
50          </Row>
51        );
52  } } export default RequestRow;
```

13.10.7 PAGES/INDEX.JS

```
1   import React, { Component } from 'react';
2   import { Card, Button } from 'semantic-ui-react';
3   import factory from '../ethereum/factory';
4   import Layout from '../components/Layout';
5   import { Link } from '../routes';
6
7   class CampaignIndex extends Component {
8     static async getInitialProps() {
9       const campaigns = await factory.methods.getDeployedCampaigns().call();
10
11      return { campaigns };
12    }
13
14    renderCampaigns() {
15      const items = this.props.campaigns.map(address => {
16        return {
17          header: address,
18          description: (
19            <Link route={`/campaigns/${address}`}>
20              <a>View Campaign</a>
21            </Link>
22          ),
23          fluid: true
```

```
24        }; });
25      return <Card.Group items={items} />;
26    }
27    render() {
28      return (
29        <Layout>
30          <div>
31            <h3>Open Campaigns</h3>
32            <Link route="/campaigns/new">
33              <a>
34                <Button
35                  floated="right"
36                  content="Create Campaign"
37                  icon="add circle"
38                  primary
39                />
40              </a>
41            </Link>
42            {this.renderCampaigns()}
43          </div>
44        </Layout>
45      ); } }
46  export default CampaignIndex;
```

13.10.8   PAGES/CAMPAIGNS/NEW.JS

```
1   import React, { Component } from 'react';
2   import { Form, Button, Input, Message } from 'semantic-ui-react';
3   import Layout from '../../components/Layout';
4   import factory from '../../ethereum/factory';
5   import web3 from '../../ethereum/web3';
6   import { Router } from '../../routes';
7   class CampaignNew extends Component {
8     state = {
9       minimumContribution: '',
10      errorMessage: '',
11      loading: false
12    };
13    onSubmit = async event => {
14      event.preventDefault();
15      this.setState({ loading: true, errorMessage: '' });
16      try {
17        const accounts = await web3.eth.getAccounts();
18        await factory.methods
19          .createCampaign(this.state.minimumContribution)
20          .send({
21            from: accounts[0]
22          });
23        Router.pushRoute('/');
24      } catch (err) {
25        this.setState({ errorMessage: err.message }); }
26      this.setState({ loading: false });
```

```
27    };
28    render() {
29      return (
30        <Layout>
31          <h3>Create a Campaign</h3>
32          <Form onSubmit={this.onSubmit} error={!!this.state.errorMessage}>
33            <Form.Field>
34              <label>Minimum Contribution</label>
35              <Input
36                label="wei"
37                labelPosition="right"
38                value={this.state.minimumContribution}
39                onChange={event =>
40                  this.setState({ minimumContribution: event.target.value })}
41              />
42            </Form.Field>
43            <Message error header="Oops!" content={this.state.errorMessage} />
44            <Button loading={this.state.loading} primary>
45              Create!
46            </Button>
47          </Form>
48        </Layout>
49      ); } }
50  export default CampaignNew;
```

13.10.9 PAGES/CAMPAIGNS/SHOW.JS

```
1   import React, { Component } from 'react';
2   import { Card, Grid, Button } from 'semantic-ui-react';
3   import Layout from '../../components/Layout';
4   import Campaign from '../../ethereum/campaign';
5   import web3 from '../../ethereum/web3';
6   import ContributeForm from '../../components/ContributeForm';
7   import { Link } from '../../routes';
8   class CampaignShow extends Component {
9     static async getInitialProps(props) {
10      const campaign = Campaign(props.query.address);
11      const summary = await campaign.methods.getSummary().call();
12      return {
13        address: props.query.address,
14        minimumContribution: summary[0],
15        balance: summary[1],
16        requestsCount: summary[2],
17        approversCount: summary[3],
18        manager: summary[4]
19      };}
20    renderCards() {
21      const {
22        balance,
23        manager,
24        minimumContribution,
25        requestsCount,
26        approversCount
27      } = this.props;
```

```
28      const items = [
29        {
30          header: manager,
31          meta: 'Address of Manager',
32          description:
33            'The manager created this campaign and can create requests to withdraw money',
34          style: { overflowWrap: 'break-word' }
35        },
36        {
37          header: minimumContribution,
38          meta: 'Minimum Contribution (wei)',
39          description:
40            'You must contribute at least this much wei to become an approver'
41        },
42        {
43          header: requestsCount,
44          meta: 'Number of Requests',
45          description:
46            'A request tries to withdraw money from the contract. Requests must be approved by approvers'
47        },
48        {
49          header: approversCount,
50          meta: 'Number of Approvers',
51          description:
52            'Number of people who have already donated to this campaign'
53        },
54        {
```

```
55          header: web3.utils.fromWei(balance, 'ether'),
56          meta: 'Campaign Balance (ether)',
57          description:
58            'The balance is how much money this campaign has left to spend.'
59        }
60      ];
61      return <Card.Group items={items} />;
62    }
63    render() {
64      return (
65        <Layout>
66          <h3>Campaign Show</h3>
67          <Grid>
68            <Grid.Row>
69              <Grid.Column width={10}>{this.renderCards()}</Grid.Column>
70              <Grid.Column width={6}>
71                <ContributeForm address={this.props.address} />
72              </Grid.Column>
73            </Grid.Row>
74            <Grid.Row>
75              <Grid.Column>
76                <Link route={`/campaigns/${this.props.address}/requests`}>
77                  <a>
78                    <Button primary>View Requests</Button>
79                  </a>
80                </Link>
81              </Grid.Column>
```

```
82            </Grid.Row>
83          </Grid>
84        </Layout>
85      );} } export default CampaignShow;
86
87
```

13.10.10   PAGES/CAMPAIGNS/REQUESTS/INDEX.JS

```
1    import React, { Component } from 'react';
2    import { Button, Table } from 'semantic-ui-react';
3    import { Link } from '../../../routes';
4    import Layout from '../../../components/Layout';
5    import Campaign from '../../../ethereum/campaign';
6    import RequestRow from '../../../components/RequestRow';
7    class RequestIndex extends Component {
8      static async getInitialProps(props) {
9        const { address } = props.query;
10       const campaign = Campaign(address);
11       const requestCount = await campaign.methods.getRequestsCount().call();
12       const approversCount = await campaign.methods.approversCount().call();
13       const requests = await Promise.all(
14         Array(parseInt(requestCount))
15         .fill()
16         .map((element, index) => {
17           return campaign.methods.requests(index).call();
18         }));
19       return { address, requests, requestCount, approversCount }; }
20     renderRows() {
21       return this.props.requests.map((request, index) => {
22         return (
23           <RequestRow
24             key={index}
25             id={index}
```

```
26             request={request}
27             address={this.props.address}
28             approversCount={this.props.approversCount}
29           />
30         ); }); }
31     render() {
32       const { Header, Row, HeaderCell, Body } = Table;
33       return (
34         <Layout>
35           <h3>Requests</h3>
36           <Link route={`/campaigns/${this.props.address}/requests/new`}>
37             <a>
38               <Button primary floated="right" style={{ marginBottom: 10 }}>
39                 Add Request
40               </Button>
41             </a>
42           </Link>
43           <Table>
44             <Header>
45               <Row>
46                 <HeaderCell>ID</HeaderCell>
47                 <HeaderCell>Description</HeaderCell>
48                 <HeaderCell>Amount</HeaderCell>
49                 <HeaderCell>Recipient</HeaderCell>
50                 <HeaderCell>Approval Count</HeaderCell>
51                 <HeaderCell>Approve</HeaderCell>
52                 <HeaderCell>Finalize</HeaderCell>
```

```
53               </Row>
54             </Header>
55             <Body>{this.renderRows()}</Body>
56           </Table>
57           <div>Found {this.props.requestCount} requests.</div>
58         </Layout>
59       ); } } export default RequestIndex;
60    |
61
```

13.10.11   PAGES/CAMPAIGNS/REQUESTS/NEW.JS

```
1    import React, { Component } from 'react';
2    import { Form, Button, Message, Input } from 'semantic-ui-react';
3    import Campaign from '../../../ethereum/campaign';
4    import web3 from '../../../ethereum/web3';
5    import { Link, Router } from '../../../routes';
6    import Layout from '../../../components/Layout';
7    class RequestNew extends Component {
8      state = {
9        value: '',
10       description: '',
11       recipient: '',
12       loading: false,
13       errorMessage: ''
14     };
15     static async getInitialProps(props) {
16       const { address } = props.query;
17       return { address };
18     }
19     onSubmit = async event => {
20       event.preventDefault();
21       const campaign = Campaign(this.props.address);
22       const { description, value, recipient } = this.state;
23       this.setState({ loading: true, errorMessage: '' });
24       try {
25         const accounts = await web3.eth.getAccounts();
26         await campaign.methods
27           .createRequest(description, web3.utils.toWei(value, 'ether'), recipient)
```

```
28            .send({ from: accounts[0] });
29          Router.pushRoute(`/campaigns/${this.props.address}/requests`);
30        } catch (err) {
31          this.setState({ errorMessage: err.message });
32        }
33        this.setState({ loading: false });
34      };
35      render() {
36        return (
37          <Layout>
38            <Link route={`/campaigns/${this.props.address}/requests`}>
39              <a>Back</a>
40            </Link>
41            <h3>Create a Request</h3>
42            <Form onSubmit={this.onSubmit} error={!!this.state.errorMessage}>
43              <Form.Field>
44                <label>Description</label>
45                <Input
46                  value={this.state.description}
47                  onChange={event =>
48                    this.setState({ description: event.target.value })}
49                />
50              </Form.Field>
51              <Form.Field>
52                <label>Value in Ether</label>
53                <Input
54                  value={this.state.value}
55                  onChange={event => this.setState({ value: event.target.value })}
56                />
57              </Form.Field>
58              <Form.Field>
59                <label>Recipient</label>
60                <Input
61                  value={this.state.recipient}
62                  onChange={event =>
63                    this.setState({ recipient: event.target.value })}
64                />
65              </Form.Field>
66              <Message error header="Oops!" content={this.state.errorMessage} />
67              <Button primary loading={this.state.loading}>
68                Create!
69              </Button>
70            </Form>
71          </Layout>
72      ); } } export default RequestNew;
```

CONCLUSION

The chapter focused on the crowdfunding platform Kickstarter and commented on its issues and in-security. This chapter discussed how these problems can be solved in an efficient way using Ethereum platform and discussed the whole development, deployment, and interaction part in a profound manner. Using the blockchain technology, the security, privacy, and the trust can be enhanced exponentially. Blockchain is a disrupting technology and in the very future it will change whole economic and commerce systems.

REFERENCES

1. L. Hornuf and A. Schwienbacher "Should securities regulation promote equity crowd-funding?" *Small Business Economics*, 49(3), 579–593 (2017).
2. (a) P. Belleflamme, T. Lambert, A. Schwienbacher "Crowdfunding: tapping the right crowd," SSRN eLibrary, 2012, https://en.wikipedia.org/wiki/Kickstarter. (b). H. Cuihong, "Research on Web3.0 Application in the Resources Integration Portal," *2012 Second International Conference on Business Computing and Global Informatization*, Shanghai, 2012, pp. 728–730.
3. M. Shi and L. Guan, "An Empirical Study of Crowdfunding Campaigns: Evidence from Jing Dong Crowdfunding Platform," *2016 13th International Conference on Service Systems and Service Management (ICSSSM)*, Kunming, 2016, pp. 1–5.
4. F. Ferreira and L. Pereira, "Success Factors in a Reward and Equity Based Crowdfunding Campaign," *2018 IEEE International Conference on Engineering, Technology and Innovation (ICE/ITMC)*, Stuttgart, 2018, pp. 1–8.
5. K. Zhang and H. Jacobsen, "Towards Dependable, Scalable, and Pervasive Distributed Ledgers with Blockchains,"*2018 IEEE 38th International Conference on Distributed Computing Systems (ICDCS)*, Vienna, 2018, pp. 1337–1346.
6. F. Hofmann, S. Wurster, E. Ron and M. Böhmecke-Schwafert, "The Immutability Concept of Blockchains and Benefits of Early Standardization,"*2017 ITU Kaleidoscope: Challenges for a Data-Driven Society (ITU K)*, Nanjing, 2017, pp. 1–8.
7. E. K. Lua and J. Crowcroft, A survey and comparison of peer-to-peer overlay network schemes. *IEEE Communications Surveys & Tutorials*, 7(2), 72–93 (2005).
8. A. Manzalini and A. Stavdas, "A Service and Knowledge Ecosystem for Telco3.0-Web3.0 Applications,"*2008 Third International Conference on Internet and Web Applications and Services*, Athens, 2008, pp. 325–329.
9. J. S. Coron, "What is cryptography?" *IEEE Security & Privacy Journal*, 12(8), 70–73 (2006).
10. Ethereum Docs. https://www.ethereum.org, 2018.
11. A. Aldweesh, M. Alharby, E. Solaiman and A. van Moorsel, "Performance Benchmarking of Smart Contracts to Assess Miner Incentives in Ethereum," *2018 14th European Dependable Computing Conference (EDCC)*, Iaşi, Romania, 2018, pp. 144–149.
12. T. Chen et al., "Understanding Ethereum via Graph Analysis," *IEEE INFOCOM 2018-IEEE Conference on Computer Communications*, Honolulu, HI, 2018, pp. 1484–1492.
13. E. Yavuz, A. K. Koç, U. C. Çabuk and G. Dalkılıç, "Towards Secure e-Voting Using Ethereum Blockchain," *2018 6th International Symposium on Digital Forensic and Security (ISDFS)*, Antalya, 2018, pp. 1–7.
14. S. R. Niya, F. Shüpfer, T. Bocek and B. Stiller, "Setting Up Flexible and Light Weight Trading with Enhanced User Privacy using Smart Contracts," *NOMS 2018-2018 IEEE/ IFIP Network Operations and Management Symposium, Taipei*, 2018, pp. 1–2.

15. B. K. Mohanta, S. S. Panda and D. Jena, "An Overview of Smart Contract and Use Cases in Blockchain Technology," *2018 9thInternational Conference on Computing, Communication and Networking Technologies (ICCCNT), Bangalore*, 2018, pp. 1–4.

16. Solidity, Docs. http://solidity.readthedocs.io/en/v0.4.24/, 2018.

17. E. Kromidha, "A Comparative Analysis of Online Crowdfunding Platforms in USA, Europe and Asia," *eChallenges e-2015 Conference, Vilnius,* 2015, pp. 1–6.

18. H. Dai, D. J. Zhang, "Prosocial goal pursuit in crowdfunding: evidence from Kickstarter", 56(3), 498–517. Article first published online: https://www.kickstarter.com/, March 28, 2019.

19. K. Williams, "Once upon a time: How a storytelling robot became a kickstarter success pipelining: Attractive programs for women," *IEEE Women in Engineering Magazine*, 9(1), 28–32 (2015).

20. L. Xuefeng and W. Zhao, "Using Crowdfunding in an Innovative Way: A Case Study from a Chinese Crowdfunding Platform," *2018 Portland International Conference on Management of Engineering and Technology (PICMET)*, Honolulu, HI, 2018, pp. 1–9.

Index